THE NORWEGIAN METHOD

THE NORWEGIAN METHOD

The Culture, Science, and Humans Behind the Groundbreaking Approach to Elite Endurance Performance

BRAD CULP

80/20 Publishing, LLC
1073 Oberland Drive
Midway, UT 84049
www.8020books.com

Distributed in the United States and Canada by Simon & Schuster

Library of Congress Control Number: 2024943392

ISBN 979-8-9892569-6-9 print
ISBN 979-8-9892569-7-6 ebook

Cover and interior design by Vicki Hopewell
Cover photo: Wagner Araujo/World Triathlon

ᚠᛟᚱ ᛗᛁᛝ ᚠᛅᚦᛁᚱ ᛅᚦ ᚠᛅᚦᛁᚱᛋ : ᚠᛅᚦᛁᚱᛋ ᛅᚦ ᚠᛁᚲᛁᚾᚲᛋ

For min far og vikingenes fedre

For my father and the fathers of Vikings

CONTENTS

INTRODUCTION

It was late spring in Bergen, Norway, and that meant it hadn't stopped raining for a few days. I was sitting in a small café having a coffee and carbohydrates with Kristian Blummenfelt, the then reigning Ironman® and Olympic triathlon champion. He was staring out the window, wondering if there would be a break in the downpour or if his upcoming easy run would be like most of the others when he's at home: wet and cold.

"We better get going—it's not going to let up," he said, looking up at the clouds surrounding Mount Fløyen, a steep crag that looms over Bergen's ancient harbor.

Earlier that morning, during a period of only light drizzle, I had asked Kristian if I could join him for his afternoon run, to which he replied, "How fit are you?"

As a perfectly mediocre endurance athlete, that was a hard question for me to answer, especially when it came from one of the fittest humans in history. I told him the truth, that I was very fit by Chicago standards but maybe not so much by his Norwegian ones. He didn't seem too concerned about pace, since it was one of the only weeks he spent in 2022 taking it easy, and I assured him I'd have no problem making it three miles up the mountain and three back down. But now he wasn't letting me off the hook.

"I don't have rain gear," I said. I was trying anything not to run in this.

He pulled out a fancy rain jacket adorned with logos from his dozens of sponsors.

"My shoes don't have any tread; I'm going to slip all over these cobblestones."

"There really aren't many cobbles," he replied.

I wasn't getting out of it. And he wasn't skipping an easy run because of what I would consider a minor monsoon. It was clear the thought never crossed his mind.

Kristian can't stand the weather in Norway—so much so that he spends as little as three weeks per year there—but it doesn't make his relationship with the outdoors any less Norwegian. The weather is what it is, and it's usually pretty bad. The trick is to keep moving—to sustain velocity—which is something Norwegians have done exceptionally well for centuries.

I expected a six-mile run up- and downhill in the rain with an Olympic gold medalist to be one of the harder workouts of my life, but I was never outside of my comfort zone during the run, and Kristian was obviously well within his. His plan called for an hour of easy running and nothing more, so I was the perfect training partner for that day.

As we finished our run back near the harbor, a group of young people walking with umbrellas shouted something in Norwegian. The day before, I had asked Kristian if he was the most famous person in Bergen, and he responded that not a single person recognizes him. I didn't believe him for a minute, and now I had proof. Surely these people were shouting at us because they recognized the newest addition to Norway's long line of legendary Olympians.

I asked them if they knew who Kristian was, and every one of them shook their heads no. They were just excited to see some people out exercising on such a miserable day and thought that warranted some encouragement. Kristian seemed quite pleased with himself in that he was both right and unidentifiable. His reasons for exercising to extremes in the elements every day go well beyond being recognized in his hometown.

I explained to the young group that this was the Olympic triathlon champion, which resulted in some more encouragement before we all got moving again. Running on the streets of Chicago in the pouring rain is a good way to get called crazy or stopped by the police. In Bergen, strangers will cheer you on as if you were on mile 26 of a marathon.

For Norwegians, getting outside and moving is a cardinal tenet of life in a way that isn't seen in many other cultures. It's part of what has made Norway far and away the most successful nation at the Winter Olympics, most notably in endurance sports like Nordic skiing and speed skating. That success has fostered a deep appreciation and ambition for sports science innovation and has created a somewhat unique method of what it takes to build an Olympic- or world-champion-caliber athlete.

The Norwegian method has gained notoriety in recent years as some of Norway's summer sports athletes have finally caught up to their winter sports brethren. (Norway has long had success in sailing, which we'll get into shortly.) Three endurance athletes in particular have thrust Norwegian training into the limelight since 2020: the aforementioned Blummenfelt; his training partner, Gustav Iden; and track superstar Jakob Ingebrigtsen. All three are very different athletes with extraordinary talent, but all three train with a similar approach.

The mileage is huge, the intensity is incredibly controlled, and no possible dataset of human biology is left unmined.

You cannot train like Kristian, Gustav, Jakob, or most of the other athletes featured in this book. That is to say, it's unlikely that you have the time, energy, or resources. And while their coaches will say that there's nothing exceptional about their talents that have made them among the best endurance athletes on earth, you probably don't have the capacity to do the sort of training that they do either. But at the core of the Norwegian method is discovering how science can become more human—more focused on the unique physiological and psychological needs of the individual—so that each human body can achieve its own peak. With that in mind, you can learn to train a little more Norwegian, which their coaches will also tell you means to just train a bit smarter and more precisely.

The Norwegian method of endurance training isn't drastically different from the programs used by the rest of the world's top athletes, and Norwegian athletes certainly aren't the only ones doing some version of it. It isn't some end-all, be-all approach that has put to bed the debate of the best way to train for endurance events. But in talking with the coaches and athletes who have popularized the method, at the core are the focus on the human, or individual, and a precision-guided training program that is both adaptable and zoomed out on the longest-term goals possible. The most effective training program is one that is individually tailored to one specific athlete—whether you're an Olympic champion or an audacious amateur.

You can tailor certain components of the Norwegian method to fit your own training program, and they can be hugely beneficial, particularly if you're at the top end of the amateur ranks. You can learn to use altitude and heat to your advantage. You can track your blood lactate levels and make your training more precise than ever. You can learn to incorporate double-threshold sessions to maximize both stress and recovery.

This book aims to analyze what the Norwegian method is and isn't, how it began, and how it's evolving alongside modern sports science. The method itself has been whipped into something of a frenzy on social media, and the athletes who subscribe to it can't win or lose a race these days without the efficacy of the method coming into question. That comes mostly from a microscopic view of what the Norwegian method is. When you zoom out, you can see that this new generation of Norwegian athletes and coaches—and the popularization of some of the training methodologies around the globe—was a long time coming and required a very specific set of circumstances that could only come from one place. We'll begin with some history, end with a lot of science, and meet some extraordinary humans along the way.

But let's begin with the thing that started it all, the single most important component in building world-class athletes: culture.

CULTURE

1

THE VIKINGS INTRODUCE THEMSELVES

When men meet foes in fight,
better is stout heart than sharp sword.

—VOLSUNGA SAGA

There have been humans in what is now Norway for at least ten thousand years, but the ones who shaped modern Scandinavian culture announced their presence with a bang in A.D. 793. That's the year most historians mark as the beginning of the Viking Age, because much of the world's history has been written by the British, and that's when Britain suffered one of the worst atrocities in its history, known as the raid on Lindisfarne. In a letter to King Aethelred of Northumbria by the writer Alcuin of York, the raid was described as such:

> We and our fathers have now lived in this fair land for nearly three hundred and fifty years, and never before has such an atrocity been seen in Britain as we have now suffered at the hands of a pagan people. Such a voyage was not thought possible. The

> church of St. Cuthbert is spattered with the blood of the priests of God, stripped of all its furnishings, exposed to the plundering of pagans—a place more sacred than any in Britain.

The "pagans" Alcuin was referring to hailed from what is now the southwest coast of Norway, and they had sailed to Britain with speeds never before seen anywhere on earth, giving the people of Lindisfarne no warning of their impending demise. Another account of the bloodshed describes the pagans as "stinging hornets," slaughtering everything in their path, including women, children, and livestock. They took particular pleasure in torturing the deacons and priests, some of which they chained together naked and drowned in the North Sea.

What all the accounts of the horror that unofficially kicked off the Viking Age have in common is an awe for the speed by which the "pagans from the north" moved over sea and land and a ghastliness for the degree of their bloodletting. In a long poem Alcuin later wrote entitled *The Destruction of Lindisfarne*, taken from *The Vikings: A History* by Robert Ferguson, he claims a "pagan warband arrived from the ends of the earth," suggesting that the attackers came from an unknown origin. But by 793, the Britons were well aware of a growing civilization up north. There are numerous historical records from a century earlier of people in the British Isles trading with reindeer farmers and whale hunters from the Arctic. There were well-established trading outposts around what are now Oslo, Copenhagen, and Stockholm, frequently visited by mainland Europeans and Britons. The only thing that changed in 793 was the speed by which their neighbors to the north moved and the violence that came with them.

The longships, however, were very real, and the origins of them can be traced back to the Hjortspring boat, which was excavated from a bog in southern Denmark in the 1920s.

The Hjortspring didn't have a sail and was instead propelled by up to 24 men and as many paddles, but it's the oldest boat humans have found that uses the same shipbuilding techniques as the longship. Built sometime between 400 and 300 B.C., the sleek and narrow ship was made to transport as many men as quickly as possible and was basically the biggest and fastest canoe humans have ever built. At 20 meters long and only 2 meters wide, it was unlike any other large ship found from that era.

Some 600 years later, when Viking society was really starting to boom, the means of propulsion switched from paddling to rowing. There is no better display of the energy and power benefits of rowing versus paddling than to watch these two Olympic sports. The earliest rowed longship discovered is the Nydam ship, which was also found in Denmark and is dated to roughly A.D. 350. At 23 meters long and 3.5 meters wide with flared sides, it was much more stable than boats like the Hjortspring, and it could transport a crew of up to 48 men pulling 24 ores.

Viking ship design served two primary goals: transport as many warriors as possible and do so at the greatest possible speed. It took them a minute to develop the power to harness the wind aboard a sleek ship, but once they did, the time they'd spent perfecting human-powered movement over water and hydrodynamics gave them a speed the world had never seen before.

It's unknown whether the Vikings developed sails on their own, but like most innovations, they were likely borrowed from somewhere.

Relative to the rest of European civilizations, the Vikings are credited as being among the earliest to sail long distances. That's certainly true, but when compared to the Egyptians and Polynesians, they were still a few millennia behind. Perhaps most remarkable about the proliferation of sailing around the globe is that the Egyptians and Polynesians developed the first sails at roughly the same time (around 3500 B.C.), but it wasn't until some 3,000 years later that it became the preferred method of movement for the Mediterranean civilizations and even longer until the pagans of the north used it to travel much greater distances over nearly any body of water.

While they may have been late to the sailing party, the Viking longship had advantages over every other sailboat that preceded it. Although no civilization will ever match the Polynesians when it came to distance, navigation, and efficiency (they could sail thousands of miles with only a handful of explorers, using essentially a catamaran), the Viking longship was designed with the ocean, sea, and rivers in mind. Whether it came to trade, exploration, or, most importantly, conquering and plundering, the Vikings aimed to do as much of it as possible over water. Their horses could travel extraordinary distances, but they were small and not ideal for warfare. More horses also meant fewer warriors per boat, so the goal was to limit overland travel as much as possible. The ability of the longship to travel rapidly and navigate (via rowing) rivers was what enabled the Vikings to conquer vast portions of what is now Russia while simultaneously taking over western European and Mediterranean civilizations via the sea.

The Vikings were also the first people other than the Polynesians to sail well beyond sight of land, excluding the many cultures surrounding the Mediterranean, which limits just how lost a ship can become (Odysseus notwithstanding). Like the Polynesians, the range of the Vikings' sailing excursions was more a factor of necessity than anything else. Unless one sought to do an arduous trek through the Arctic Circle, the only way to find out what was south and east of the Vikings' known world involved a long trip over rough seas—and they had the hardened people to do it.

Iterating on intelligent design

By the time the warring Viking society had grown to the point that world domination was the next logical step, the design of the longship had been perfected enough to make that possible at an extraordinary rate. The best and most important ships had sails made of wool, which was somewhat unique for the time. They were heavier than sails made from woven plant fibers, and they involved a lot of work to turn into proper sails, but once they were, they were unmatched. They had incredible tensile strength to harness and hold as much of the wind as possible, and they were resistant to ripping or rotting.

Powerful sails on sleek boats present a number of challenges—namely, keeping the boat upright. Overcoming these challenges involved a delicate balance of sand and stone ballasts, men, sail size, keel, rudder depth, and probably a few more things that the Vikingeskibs Museet (Viking Ship Museum) in Oslo has yet to figure out. These were not easy ships to sail, and nearly every voyage required a unique setup. What we do know is that it was specifically the keel of the longship that made it a perfect raiding vehicle. It typically sat just one meter deep, making some of the shallowest inland waterways navigable. That was on the very same boat that could carry 40 to 50 men at speeds up to 15 knots (roughly 17 mph) with only a single sail. Raiding coastal towns was one thing. Everyone was doing that in the first few centuries Anno Domini. But the ability to plunder the coast and any villages that lay upstream made escape impossible and surrender (likely death) inevitable.

The old Greek, Roman, and Carthaginian ways of conquering were out. They used ships that were as long as 150 feet, with deep hauls, hundreds of ores, and often multiple levels. These were basically floating towns, and most couldn't be beached for rapid disembarkment. A specific design was required to make way through rivers when the water was deep enough. The lightweight and easy maneuverability of the longship gave the Vikings a new, more efficient means of conquering. Perhaps best of all was that there wasn't a massive, tall ship in the distance warning coastal inhabitants of incoming danger—except for the people of Lindisfarne. The Viking method was so simple and effective that a single warring party could pillage multiple towns on the same day: reach the coast without announcing arrival, pull the feathery ship on the beach for a bit, plunder, carry or row the boat to the mouth of the river, repeat the process upstream before the target has any idea what's coming.

The Vikings had more than just awesome boats on their side. The Roman Empire was in shambles, and parts were being taken over on a first-come, first-serve basis. That led to disorder throughout Europe and western Asia, as warring clans sought to get the upper hand on the land those loyal to Rome were leaving behind. A large swath of the world was ripe for a takeover, but that conquest came from the most unexpected of places and an empire that wasn't built to last.

The end of an era of conquest

Much like the beginning of the era, the end of the Viking Age that most historians agree on has been decided by the British. It typically coincides with the Norman Conquest of 1066, when the Duke of Normandy—later known as William the Conqueror—defeated (and killed) Norwegian king Harald Hardrada at the Battle of Hastings, freeing Britain from 273 years of Viking rule. In a cruel twist of fate, the Vikings were mostly responsible for the creation of Norman society (thus the name). The Francs, who were also Germanic in origin and had come to rule most of modern-day France, offered the Vikings rule over the surly people of their north, as long as the Vikings would stop trying to seize their already beloved Paris. They agreed, but they never forced the Normans to take their language and didn't do a great job keeping watch, so it only took a few generations for a separate society and, ultimately, mutiny.

Fewer than three centuries is relatively short as far as empires go, but the reach was impressive in such a short period of time, and the effects on both Europe and Scandinavia remain to this day. The empire stretched west to Greenland and likely parts of modern-day Canada, east to what is now central Russia, and south to include nearly all of coastal Spain and Italy. The Byzantines still had their hold on most of the eastern Mediterranean, but both the Black and Caspian Seas were firmly under Viking rule by the end of their run thanks to their ability to traverse the Volga River.

Unfortunately, they didn't spend any time raiding the river systems of central Europe, and that would lead to a rather abrupt downfall at the hands of the Normans, Slavs, and Francs, who were largely spared from Viking rule. But nearly three centuries of raids had made them rich, and it was during this time that Scandinavian—and Norwegian—culture began to really take shape. The cities that are now Oslo, Bergen, Copenhagen, Stockholm, and Götborg all exploded in population and wealth during the Viking Age, and even after the fall of the empire, they remained some of the most important ports in the world.

As much as the centuries of raiding left their mark on continental Europe and western Asia (i.e., by decreasing the population a bit), they had an even bigger impact on Scandinavian culture. The cultural influences of the British and Celtic Isles were particularly profound, with Norway becoming fully unified as a Christian nation under King Olav the Second in the early 11th century. The thought of Norway being a Christian nation would have many Viking jarls (chiefs) rolling in their spacious tombs, but luckily for them, it wouldn't last long. While the Church of Norway officially claims 68 percent of the country's 5.4 million people today, that's only due to an ancient law that decrees that anyone born with at least one parent member of the church is forever a member of the church. In reality, modern Norway is among the most secular nations on earth, with only 2 percent of the population regularly attending services. The pagans of the north have returned to their ancient ways.

Norway's formative years as a Christian monarchy involved a lot of turmoil, with a handful of rulers trading places and raids from displaced Germanic tribes east and south providing a bit of karma for three centuries of Viking mischief. Without any clear laws of succession, Norway became largely divided back into tribes tied to their Germanic ancestry and actually endured one of the longer civil wars in history, with ongoing conflict and 12 alleged kings from 1120 through 1240. After enough of them had died off, 13-year-old Haakon

IV Haakonsson was decreed king by the church and at least part of a rudimentary parliament in 1217, and when he married one of his remaining opponent's daughters eight years later, unification of Norway as its own country was nearly complete for the very first time. All that was left to do was kill off the remaining person claiming to be the rightful heir of Harald Fairhair, the first to be given the title of king of Norway in 872. That false king—as Norwegian history is often written—was Skule Bårdsson, who had been given the title of duke by the young King Haakon, but that was deemed insufficient enough that he launched an unsuccessful coup in 1240 and was killed.

While the Vikings, as we've known them throughout the last chapter and a half, ceased to exist, their way of life is ingrained in Norwegian society, particularly their impulse to explore the world via the sea—and to do it faster than anyone else.

A longship revival and the dawn of an Olympic dream

Perhaps the most impressive longship to be unearthed was discovered in 1880 in southeastern Norway. The Gokstad is believed to have been built around 990, relatively late in the Viking Age. It was found fully intact, including even the decorative dragon on the prow that was there to strike even more fear into their soon-to-be victims. The ship struck so much awe in Captain Magnus Andersen, a Norwegian sea captain and explorer, that he commissioned an exact replica to be built so that he could attempt to sail across the Atlantic with it, paying homage to the maritime genius of his ancestors.

With a keel of just one meter, Andersen and 10 others left Bergen on April 30, 1893, and arrived in New London, Connecticut, 44 days later to much fanfare. But that wasn't quite enough for Andersen, who was well versed in Viking history and determined to remind the world why they were once referred to as the "river kings." With the World's Columbian Exposition—an event intending to commemorate Columbus discovering North America, which many Norwegians take issue with until this day—taking place in Chicago a few weeks later, Andersen decided to take his Gokstad replica up the Hudson River, down the Erie Canal, and across four of the five Great Lakes to reach Chicago, and when he arrived, he was met with even greater fanfare. After stealing the show at the World's Columbian Exposition, Andersen took his longship down the Mississippi to spend the winter in New Orleans. In total, Andersen and his crew sailed and rowed their ship approximately 4,800 miles in 1893. After nearly a century of living in a basement of Chicago's Lincoln Park Zoo, the Gokstad replica was moved to various locations around Illinois's Fox Valley thanks to the insistence of a (relatively) large Norwegian and Swedish community. The prowl of the ship—a dragon named "Freya"—is on display at the Geneva Public Library (where much of this book was written), and the ship itself is open for viewing at Geneva's Good Templar Park, where it's locally referred to as the "Viking Ship."

Andersen's feat was worldwide news, and it rekindled Norway's zeal for sailing at an ideal time. Just three years after his Atlantic and North American odyssey, the very first Olympics would take place in Athens. Unfortunately, sailing wasn't contested due to rough seas, and six of the eight members of the Norwegian delegation were there for shooting, not sailing. Norway would have to wait until the 1908 Rome Olympics to make its sailing debut. In the 1908 Games, Norway had just four men on an eight-meter boat and finished fourth. Four years later, Norway won two of the three classes—much to the chagrin of the

Swedish hosts. After World War I caused the cancellation of the 1916 Games, sailing was the marquee sport of the 1920 Antwerp Olympics, at least in terms of the number of events, with 14 classes racing in the North Sea. It remains the largest sailing competition ever organized at the Olympic Games. The Norwegian Sailing Team turned in one of the most dominant performances in Olympic history, winning 7 of the 14 classes and medaling in 11 of them. A closer look at the results shows that Norway was the only country entered in six of these races, making their gold medals akin to being voted dog catcher in an uncontested small-town election (two additional races were canceled when even Norway failed to muster a boat), but this in itself speaks to how much more seriously the Norwegians took competitive sailing (or "yachting," as it was called back then) than did the rest of the world.

The Norwegian Sailing Team remained the most dominant in the world until the 1960s, when the Cold War pushed the United States, Russia, and the two Germanys to supremacy across nearly every Olympic sport with often nefarious methods. But the unprecedented success of the 1920 team meant a lot for a nation once hell-bent on taking over the world, and once again, the feats of Norwegian sailors came at a rather opportune time.

The Summer Olympic Games had grown from a rather local affair in 1896 to the most important thing in the world after World War I, so the juvenile International Olympic Committee (IOC) decided to host the first Winter Olympiad in 1924 in Chamonix, France, and it was Norway's chance to once again take over the world—though this time, of sports. They ruled both the downhill and Nordic skiing events to claim 17 medals, 4 of which were gold. The Norwegian contingent was led by Thorleif Haug, who won gold in all three cross-country skiing

Summer Olympics: Norway's Medal Count

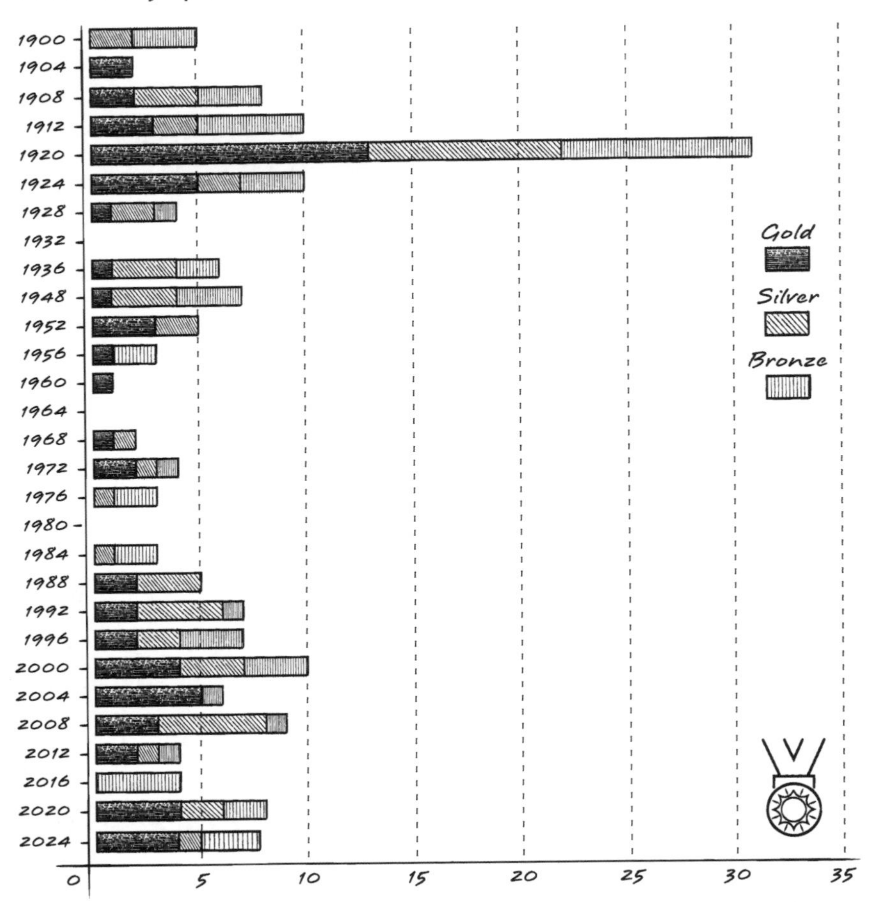

Norway was ranked 6th in the overall medal count in the 1920 Games, amassing 31 medals, most of which were in sailing. In more recent years, Norway is lucky to break into the top 20 countries.

distances as well as bronze in the ski jump. (Fifty years later, a calculation error revealed that he had actually finished fourth, and his daughter posthumously returned his medal.) Finland claimed 11 medals, 4 of which were also gold, stoking a sporting rivalry that burns strong to this day. No other nation came close to claiming double-digit medals.

HUMAN

3

INGRID KRISTIANSEN: THE MOTHER OF THE METHOD

A wise [wo]man does all things in moderation.

—GISLI SURSSON'S SAGA

It's fitting that the current, golden, and mostly male generation of Norwegian athletes was fostered by a pair of women who were the first to truly dominate running. When Kristian Blummenfelt announced after the 2021 Olympics that he intended to win the Ironman World Championship, he was largely mocked online because what he was attempting to do was beyond unconventional. That was not how the sport had worked since its inception: You could not be the best in the world at two events that were so varied at once; that was simply against the rules. Kristian, of course, proved that the old rules were stupid and needed to be updated. But he was hardly the first Norwegian to do it.

One for the record books

The first person to create an association between Norway and endurance sports dominance on a global stage was Grete Waitz, who became both mentor and rival to the woman who would become Norway's longest-standing world record holder. Born in Oslo in 1953, Grete became a superstar at home after winning the 1978 World Cross Country Championship but went unrecognized in Manhattan later that year as she took in the local attractions with her husband, Jack, the day before the New York City Marathon. Race director Fred Lebow had accepted her application to join the women's elite field over a colleague's objections, despite the fact that Grete had never run farther than 12 miles in training, let alone raced a marathon. It was his hope that she'd serve as an unwitting rabbit—or pacer—for Christa Vahlensieck, who had lowered the world record to 2:34:48 the year before and gone public with a goal of breaking the 2:30 barrier in New York City. Grete had plenty of time for sightseeing on race eve because although Lebow had covered her airfare, he had not invited her to any press events.

Conditions were close to perfect when the race set off at 9:00 a.m. from Fort Wadsworth on Staten Island, with cool, dry air stirred by a light breeze out of the northwest. Grete wore pigtails, a blue-and-white Norwegian singlet, and race number 1173F. Thwarting Lebow's expectation of self-sacrificial pacemaking, she shadowed the favorites, though not without misgivings.

"I was used to running so much faster," she said later. "It felt so slow."

As the miles clicked past, Grete became increasingly restless, at times nearly clipping the heels of the plodding world record holder in front of her. Her patience ran out on the long climb up the Queensboro Bridge, where Grete rocketed ahead of the erstwhile leader and proceeded to blast out a second-half split of 1:13:59 to take the win in 2:32:39.

Needless to say, Grete Waitz never wore the number 1173 again. But she did win the New York City Marathon again—eight times. Today, however, she is regarded as only the second-greatest Norwegian distance runner of her era.

Star-crossed runners steeped in a similar method

The now indissociable names of Grete Waitz and Ingrid Kristiansen were first paired at the 1971 European Track and Field Championships, where a 17-year-old Grete (then known as Grete Anderson) represented Norway in the 1500 meters alongside 15-year-old prodigy Ingrid (then known by her maiden name of Christiansen), who was Norway's most promising cross-country skiing talent, dabbling in running to broaden her fitness base. It was an extraordinary coaching inclination and decades before the Norwegian Children's Rights in Sports Act suggested maybe kids should be doing just that. Unfortunately, it wasn't such a memorable meet for the two young Norwegians. Grete finished eighth in her heat and failed to advance, and Ingrid—the youngest competitor in the field—got shoved off the track and failed to finish. The event was so jarring for Ingrid that she went back to skiing, becoming one of the best 15-year-old skiers on the planet, while Grete remained on the path to becoming one of the premier distance runners on earth. It appeared that the two athletes, who were roommates at that event, might never cross paths again.

By 1976, Ingrid was on her way to the Winter Olympics (albeit as an alternate), and Grete had made her way to the top of the world on the track. Running was still a major part of Ingrid's ski training, so she decided to enter her first marathon the following year and finished in 2:45, which was less than 10 minutes off the world record at the time. The talent that she always thought was there was greater than she knew, and she felt confident that she had an even greater aerobic base than most—maybe all—of the top marathoners at the time. Admittedly, she had a worthy rival in Grete, who was about to reset the marathon world record in each of her first two attempts at the distance.

From the outside, the two women in their mid-20s were approaching the marathon from very different places. But as far as their bodies were concerned, they had remarkably similar preparation. Both were doing extremely high volume from a very young age, with Ingrid perhaps having the advantage of doing a low-impact activity to ramp up her aerobic base. It's something she still ponders today, both when thinking about her own career and when observing promising young runners.

"Maybe coming from skiing was my big advantage," she speculates when I ask what made her so much different from the other women marathoners of her day. "You see, no women were coming from distance running back then. Girls didn't run long distances—especially in Norway."

Ingrid was born in 1956 in Trondheim, a port city 250 miles north of Oslo, not far from where she calls home now. She had one brother four years older, and their parents had no particular sporting interests, letting the kids fill their days with whatever activities they liked. There weren't many organized sports for girls in northern Norway in the 1960s, so Ingrid took part in whatever activities her brother and his friends were up to, which were usually skiing, running, or soccer. By remaining mostly uninvolved in their daughter's sporting life, Ingrid's parents had inadvertently created the perfect upbringing for one of the greatest runners in history.

By age 12, her skiing talent was obvious, and even a girl from way up in Trondheim couldn't be ignored, so she was placed in the local ski club's junior elite program. That involved a couple of easy runs each week, and she was able to develop her running alongside her skiing, albeit at a slower pace. She speaks fondly of her time as a young skier—which is now nearly 50 years ago—chasing the boys through the forests and oftentimes beating them.

It might come as no surprise that the type of training these young Norwegian skiers were doing in the '60s and '70s looks similar to the contemporary method that has been popularized by Norway's summer sport success.

"It was very much 80/20 training," she explains. "Eighty [percent] easy and long, twenty [percent] hard and short. But never harder than race threshold in training."

"Ever?" I ask her to clarify.

"Never. Of course not. I would never have done that."

She replies so matter-of-factly that it's like she's surprised by the question. Like, why would I even think to ask? It's clearly not something they ever thought about doing, and it stuck with her throughout her career.

A world-class state of mind

Ingrid continued skiing and running throughout the '70s, which were her high school and college years. There was not yet a *toppidrett skole*, or elite sports high school—which we'll get to shortly—to develop athletic talent, even though she was among the best in the country at both sports.

In a 1986 issue of *Sports Illustrated*, Grete recalled her first time meeting Ingrid, at the European Championships in 1971. “I was amazed to see Ingrid morning and evening doing push-ups—lots of them. That was for skiing. Running was obviously secondary to her.”

What Grete didn’t understand back then, and perhaps not ever, was that running and skiing would always be secondary for Ingrid. Even while pursuing her Olympic dreams on skis and chasing Grete on the track, she earned a medical engineering degree and began working at the Norwegian Cancer Research Institute in Trondheim. It was a quiet job, as she describes it, and a much-needed opportunity for her mind to be occupied by something other than running or skiing.

“I couldn’t train more than two hours a day, so the rest of the day, what can we do with that?” she says. “In the ’70s and ’80s, we weren’t trying to make it easier. Even before I was a mom, I was working full time, but still, it’s a part of who you are. Too many kids now are trying to be full-time runners or full-time [athletes]. It can work for very few, but most need to fill their life with more.”

On top of her career at the institute, she began crocheting and knitting regularly, has kept at it for 40 years, and, expectedly, is very good at both of her more relaxing pursuits. But she always pushed herself hard in training, and she still needed to get past Grete, whom she had grown accustomed to finishing behind. They would race a handful of times each year at 1500 or 3000 meters throughout the ’70s (there was not yet a 5000 for women), and Ingrid finished second every time, without making it particularly close. Somewhere, in a closet in Trondheim, she estimates that there are 20 to 30 silver medals from the Norwegian championships. The corresponding golds all belong to Grete, who passed away in 2011 at age 57 after a lengthy fight with cancer.

The relationship between the two Norwegian and world greats was fluid and filled with emotion throughout their respectively impressive careers. Ingrid developed something of an inferiority complex that became so powerful that she sought out a sports psychologist to help her believe that she could be the one winning races. Norway was at the cutting-edge of sports science, but in the late 1970s, sports psychology was a foreign concept, although psychology certainly was not. In such a small country, Norway's best athlete had no problem connecting with Norway's best psychologist, Willi Railo, who had gained local fame with businesspeople and musicians. Railo taught Ingrid the power of visualization, using cassette tapes that would talk her through specific instances in which she controlled the race to beat Grete. Talking to someone other than her husband and coach gave her a fresh perspective on why she was racing, but there were more setbacks ahead before the big breakthrough.

A stress fracture kept Ingrid out of the 1978 New York City Marathon, where Grete set a world record of 2:32:30 in her first attempt at 26.2 miles. Ingrid had run her first marathon a year earlier in London, finishing in 2:45. At the time, she was still training for what would be her final cross-country skiing world championships and had never run on pavement in her life. Prior to the marathon event, she only ran a couple times a week, on trails, at a very easy pace. And never more than an hour. A 2:45 result on the heels of such passive training was all she needed to know that being the best distance runner in the world could be in her future. But first she needed to become the best distance runner in Norway.

It helps to have someone blaze a trail for you, but Ingrid had been second long enough. "She was not my ideal, in the sense that I wanted to imitate everything about her," she told *Sports Illustrated* in 1986.

"But she showed what women could do if we trained like men. She showed that Norwegian girls can be the best. It was frustrating to be behind her for those years, but there was always the feeling, 'If she can do it, I can do it, too.'"

The debut of the professional-runner mother

In the '80s, while training to be the best, Ingrid experienced a most unexpected circumstance. At that time, Ingrid wasn't just playing second fiddle to her countrywoman. Women's distance running was experiencing its first boom, and that meant real competition from all over the world, most notably in the United States. On the track, it was Mary Slaney, and on the road, it was Joan Benoit, who would trade off the world record with Grete throughout the early '80s, bringing the target down to 2:22:43 by 1983.

That same year, Ingrid won the Houston Marathon in 2:33:27, and it was one of the most disappointing results of her career—at least for a few weeks. It was supposed to be her first time running under 2:30, and everything in her training pointed to something well under that. After a rough month of training following Houston, she took a pregnancy test at the behest of her coach, which was positive, and the next day, she found out she was four months pregnant. Missing her period was normal throughout her career, so she hadn't thought anything of it. It was a shock mostly because she didn't believe she could become pregnant: "I don't think before Gaute [her son] that she really believed she could be a mother," her husband, Arve, told *Sports Illustrated*. "She thought her whole hormone system was not working in that way." But this news shifted something inside her as both a human and an athlete: "Now she is a more fulfilled person," Arve said. "There is nothing missing."

She continued to train somewhat regularly throughout her pregnancy, determined to come back stronger than everyone, even though most assumed her career was over. In 1983, there weren't many examples of women returning to elite sports after giving birth, certainly not in distance running.

Even more than 40 years later, she remembers never having a doubt that she'd return to running at the highest level. "I just looked at the three best male runners in Norway at the time. They had one, two, and three kids, so I thought, 'Why should I stop running to be a mom when the guys didn't?' I just got back to a normal life after becoming a mom. There was working, there was running, and there was family."

Nearly five months to the day after Gaute was born, Ingrid was back in Houston, defending her title with her first sub-2:30 run—2:27:51. Less than five months later, in London, she set a partial world record of 2:24:26, which was disputed because Joan Benoit had run two minutes faster in Boston the previous spring. But seeing as how it was a point-to-point course, it wasn't certified. The following year, Ingrid returned to London and resolved the dispute, running 2:21:06.

A triple threat: The 5000, the 10,000, and the marathon

Perhaps Ingrid's most impressive feat occurred just three months after her marathon world record, when she smashed her own world record in the 5000m on the track, running 14:37. She had become the first woman to run under 15 minutes just one year earlier and 11 months after giving birth. She also became the first woman to run under 31 minutes in the 10,000 in 1984 before smashing her own mark in '85 with a 30:13 run in front of a raucous crowd in Oslo. That record stood

CULTURE

4

A NATURAL PIPELINE FOR ATHLETIC DEVELOPMENT

Few are bold in old age that are cowardly in childhood.

—VOLSUNGA SAGA

Norway's aptly named Olympiatoppen is an organization that is somewhat unique in world sport. It's a joint project between the Norwegian Olympic and Paralympic Committee and the Norwegian School of Sport Sciences, where a handful of elite coaches handpick athletes with true medal potential from various sports and try to set them on the path to gold. The head of the project is a man named Tore Ovrebo, a former Olympic rower who has been exceptional at finding and developing gold medal talent in Nordic skiing and speed skating.

At a press conference a few days before the start of the 2022 Beijing Winter Olympics, Ovrebo was asked how many medals he expected the Norge to win. He replied "32," which caught the reporter by so much surprise that he had to clarify that he hadn't misheard Ovrebo's sturdy Norwegian accent.

Winter Olympics: Norway's Medals by Sport

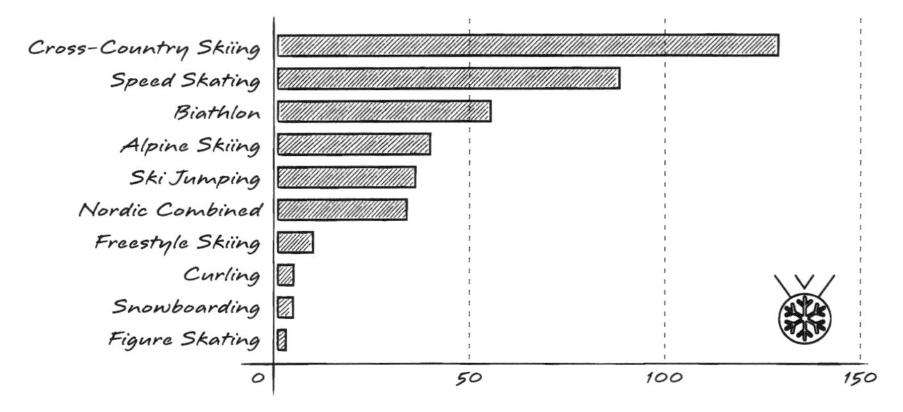

Norway's love for the outdoor life and skiing are clear in the dominant performance in cross-country skiing. Speed skating has also brought in steady medals, but in more recent decades Norway's athletes have won 84 medals in biathlon and alpine skiing.

youth to slide down tubes of ice either feetfirst, headfirst, or aboard tiny missiles with a few of their friends, which is something Norwegians have had too much common sense to do at a high level. But even Germany and Austria haven't gone to the lengths Norway has when it comes to getting kids in an elite pipeline as early as possible and keeping them in it. While it may sound like a rigid and strict upbringing for Norway's very best and brightest skiers, the secret to the sustained success is that it's anything but.

Playing the long game with kids in sports

The Norwegian Olympic and Paralympic Committee adopted the uniquely Norwegian Children's Rights in Sports Act in 1987. The 12-page document regulates sports for kids up to age 12, with the sole focus on keeping their experience positive and fun. There are limits as to how many kilometers kids can travel for competition, how often they can compete, and how awards, rankings, and results are published. Soccer (*fotball*) is still the most popular sport in Norway, ahead of Nordic skiing, but there are no lists ranking the best 10-year-olds in Oslo. The little bit of sports media that exists in Norway does not waste time debating which preteen may be the next Erling Haaland. They leave that to the British press.

Needless to say, the experience of Norwegian youth in sports is vastly different from that of many American children, where competition is paramount and finishing atop the rankings is almost always the end goal. Two separate organizations in the United States rank kids in basketball down to first grade—that's six years old.

The mission statement of the Children's Rights in Sports Act stands in stark contrast to the way youth sports are treated in other sports-crazed nations, and it's just a beautifully matter-of-fact way to approach kids in sports: "Children are children and not small adults. All sporting activities and competitions should be open to any child wishing to participate. Furthermore, such activities should be developed and adapted for children so that they are encouraged to learn new things and are motivated to continue playing sports for as long as possible."

It's more than just words. It's been playing out at the Winter Olympics for the past 40 years, and now many of the warmer-weather athletes are catching up using the same development system and living by the same credo. The focus for youth is on skills development and fun. Teach kids perfect mechanics while letting them play with their friends. Let the competition come naturally. These are the descendants of Vikings, after all.

Kajsa Vickhoff Lie is one of the top alpine ski racers in the world and made her World Cup debut at age 18. When asked about the unending success of Norwegian ski academies by *Ski Racing Magazine*, she simply said, "We get as many [girls and boys] as possible to continue ski racing as long as possible." She went on to highlight the importance of friendship in her own development and passion for the sport. Nevertheless, girls stayed in the same club throughout their entire youth because that's where their best friends were. That was their place of play. By the time she was 16, she had all the skills necessary for the next phase of elite development, and Norway, of course, had something uniquely Norwegian to offer.

The advent of elite sport schools

Strewn throughout Norway are 16 toppidrett skole that kids and their parents can apply for beginning at age 16. Many are a collaboration between the local school council and Olympiatoppen, with the goal of bringing a better school-sport balance so that teenagers can excel at both. In reality, it also maximizes the number of hours per week that they can train for their sport while still getting a good education. It is Norway, after all.

Like Olympiatoppen, these schools were started shortly after the announcement that Norway would host a Winter Olympics in 1994, in hopes that their home Olympic Games would go much better than the previous three. The first eight academies were focused mainly on Nordic skiing or skating, but now the best ones in Oslo and Bergen focus on developing athletes in everything from Tae Kwon Do to triathlon.

Each of the 16 schools is limited to only a handful of athletes for each sport, and it's the best route to national- and world-level competition, so getting in is extremely competitive. It's so competitive that Gustav Iden, who would go on to become a triathlon world champion at age 23, was rejected at age 16 from the triathlon program at Toppidrettslinjen, the top school for summer sports in western Norway. He would have to wait a year to join Kristian at the school's new triathlon program, and the rest is history that is still being written.

The schools have caused some fervor in Norway because many are privately funded and require parents to pay tuition—something that isn't exactly popular in a place that has been one of the leaders in socialized education. Of course, there are scholarship opportunities and incentives from the coaches and leaders of each school to get the absolute most talented teenagers as student-athletes. It creates a case of the haves and have-nots, although those who aren't fortunate enough to gain acceptance into one of these schools still have the fall-back option of one of the best public education systems on earth. For those young athletes with the talent and resources, toppidrett skole is a dream come true.

"You don't feel like you're choosing between your sport or your education. You're allowed to fully commit to your sport, and the education fits into that schedule—not the other way around," explains Siren Seiler, who ran track at Norges Toppidrettsgymnas (NTG), which roughly translates to "Norwegian Elite Sports School." Located in the small town of Kongsvinger, about 50 miles northeast of Oslo, it's where many of the not-so-big city's best young athletes attend school. "The big thing for me was being exposed to athletes from all over Norway doing all different sports. It was good to be surrounded by kids with similar goals but not have this tunnel vision on running."

This approach to advancing the student-athlete is similar to the dozen or so elite basketball academies in the United States, which provide the absolute best coaches and maximum playing time along with enough of an education to call it high school. Montverde Academy in Florida is perhaps the most notable and has nurtured the talents of players like Joel Embiid, R. J. Barrett, and Ben Simmons in recent years. Montverde also recruits and trains elite youngsters who excel in tennis and soccer and in recent years has nourished some of America's top talent on the track. Norway's sports school program encompasses a wider variety of sports, seemingly handpicking unique ones with the intention of developing a world champion. While some of these schools, like NTG, run on paid tuition, most enjoy plenty of government support and merely serve as a bridge until the time when the very best student-athletes can join the national team and receive the full financial support they need. It's distinctly different from Germany or Austria, where kids as young as 14 or 15 are given national team status and funding—well before they debut on an international level. Perhaps most remarkable about Norway's dominance is that their national Nordic skiing teams are so much smaller, and most of their athletes are mostly paying their own way until age 18 or later.

That's changing, and more fringe sports in Norway and around the world are becoming more professional. Toppidrettslinjen, which produced both Kristian and Gustav, was founded to train swimmers, footballers, and, of course, skiers. The triathlon program—and subsequently the Norwegian Triathlon Federation—came somewhat by happenstance. Kristian was a good but not great swimmer when he entered Bergen's first triathlon at age 14. There were fewer than 100 people, and most of them were at least three times his age. He won the sprint-distance race a few minutes clear of any competition despite doing basically no bike or run preparation. With a potential triathlon prodigy in their hometown, the school enlisted Kristian as the first

With only 15 athletes to mark the start of Triathlon Norway's elite program in 2011, there was no need for a weeding-out process; they needed to build it from the ground up. After 10 years of building, 3 of Norway's original 15 athletes were on the start line of the Tokyo 2020 Olympics (which took place in 2021), and they went on to finish 1st, 8th, and 11th.

Like most coaches in Norway, Arild is quick to dismiss talent as just getting lucky with the gene pool. "Of course we had a couple of gifted athletes, but there are gifted athletes everywhere. We're a small country, and we learn to work with what we have. I firmly believe that if you have the right system, you can more or less get elite results with what you have available."

It's why Arild is confident Norway's moment at the top of triathlon can and will be more than momentary. He believes that getting talented athletes into the right system is more important than finding the most talented athletes possible. And because Kristian and Gustav moved on to work exclusively with Olav Aleksander Bu after their illustrious Olympic debut and focused on longer distances, Arild finds himself with another unexpected opportunity. He has been left in charge of finding and building the next great Norwegian triathlon champion.

He's going about that the same way he did when he first got started: to learn from what other Norwegian coaches are doing and find the simplest way to apply it to triathlon training. Arild now works in partnership with the Norwegian Rowing Federation, which is fitting because so much of his own application of the methodology is based on Sæterdal's study of rowers. It may seem counterintuitive to have your very best coaches stretching their already thin time across two sports, but there isn't often a choice in a country so small, and that cross-pollination has been a net positive for all sports: "The coaching environment is very small. You talk to them, you learn from them, you exchange different experiences. We learn from each other. That's the way it's always worked. I like the way we work in Norway."

Success, the low and slow way

Whether he's borrowing from Norway's best rowing, running, swimming, skating, or skiing coaches and scientists, Arild's takeaway is almost always the same: intensity control over all else. It's the only way to get in the volume necessary to build a world champion over a long period of time. Arild was one of 12 Norwegian coaches who participated in a 2024 research study to pinpoint just how much of their training was spent at low versus high intensity. The researchers also asked the burning question, Just how low is low?

The training for each endurance sport in the study was analyzed according to a unique, six-zone system rooted in intensity. Six was deemed a fair meeting ground between the five-zone system that's common with runners and the seven- (or eight-) zone system used by most Norwegian winter athletes.

The study found that elite Norwegian athletes from the six sports studied spend at least 80 percent and as much as 90 percent of their time in zone 1—with triathletes, runners, and cyclists being closer to 90 and rowers, skiers, and skaters closer to 80. A big reason for this is the added crosstraining and plyometric training done by winter sports athletes. The data Arild contributed to the study included a lot of his early work with Kristian and Gustav, where they were training an average of 25 hours per week from the time they started working together, with nearly half of that time spent in the pool. It was both where they could do the most volume in terms of time and where they could inject a bit of intensity without any cost to the other two disciplines. The remaining hours were split relatively evenly, with slightly more riding early in the season and more running later on. It's considerably fewer hours than what they're doing today. A 25-hour work week with half of that time spent in the pool is what the easiest weeks of their year look like, but to get to the 35 hours of work they can consistently do as grown men, they had to start high, and that meant a lot of time in the upper reaches of zone 1 or lower parts of zone 2, which is just to say: very low.

For the researchers conducting the study and the 12 coaches involved, lactate was the key metric used in keeping the low sufficiently low—and it was what Arild began using with his triathletes almost from day one. But he's quick to point out that it's not always the best tool for new athletes. The goal for both new and world-class athletes should be to get to the point where they can listen to their own bodies and control the intensity from within, because more control means more volume, and more volume almost always leads to a fitter human.

He may now be known as the coach who first developed two of the best endurance athletes on earth and the leader of one of the most successful Olympic federations of the century, but he's still spent more of his career coaching amateurs than elites. He's quick to point out that most of what he's learned developing Olympians can trickle down to any ordinary endurance athlete, particularly in triathlon, where he sees so many doing it so wrong, mostly by making their easy training just a bit too hard.

“If you’re serious about your training, I really think lactate is the best tool to learn how low is low and how to listen to your body. But it’s expensive and invasive. And for some athletes, if you tell them you have to use this certain tool, it will turn them away from a certain type of training. You can do plenty of good training with just heart rate.”

SCIENCE

6

TRAINING PRINCIPLES OF THE NORWEGIAN METHOD

Let no man glory in the greatness of his mind,
but rather keep watch over his wits.

—HÁVAMÁL

The simplest way to summarize the Norwegian method is likely something you've read or heard before: It's a very high-volume, relatively low-intensity approach to training, with threshold sessions controlled by lactate measurement. This remains a good way of describing its origins at the turn of the century. But like any good training method, it's been ever evolving since.

The element that has garnered the most attention over the last few Olympic cycles has been the use of double-threshold days, which means doing two lactate-controlled threshold workouts on the same day. In theory. In practice, some of the very best athletes who adhere to the Norwegian method have a wide range of what they consider a threshold workout, and some of them have gotten so good at perceiving their own rate of exertion versus exhaustion that they can do away with the lactate meter most of the time.

The logic behind a double-threshold day is that you can maximize the aerobic benefit and recovery all at once by doing two hard sessions on a single day instead of spreading them throughout the week. It defies a lot of conventional endurance training wisdom that claims hard workouts (or days) must be followed by easy ones. That's been particularly true in running, where the consequences of gravity are so much greater and the hard/easy rule has guided coaches and runners for more than a century. But part of the reason it's worked so well for Norwegian endurance athletes—and now athletes from all over the world—is because it may be a smarter way of stacking the hard and easy workouts to provide both maximum stimulus and adaptation.

It also eschews any notion that the Norwegian method is somehow easier than more traditional methods. Whether it's two runs, rides, swims, skis, or any mix an athlete can come up with, nailing two hard sessions in a single day means committing to a certain level of pain, especially for the second workout, which is begun in an already fatigued state. When done right (i.e., with proper intensity control), doing a threshold workout with fatigued legs (or arms or both) can provide a much greater response than doing a single hard workout, and then that response can be amplified with the added time between hard days.

Mind the fine line between zones

In the purest form of the Norwegian method, a portable blood lactate meter—once considered an exotic device only for Olympians and now available for less than $200 on Amazon—is central. Its purpose is to keep the athlete from going too hard in either workout and effectively dipping into the well in such a way that it would need to be refilled the following day (or days). A further nuance of the Norwegian method—in its purest

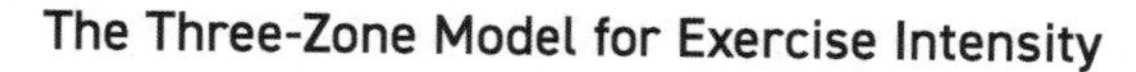

The Three-Zone Model for Exercise Intensity

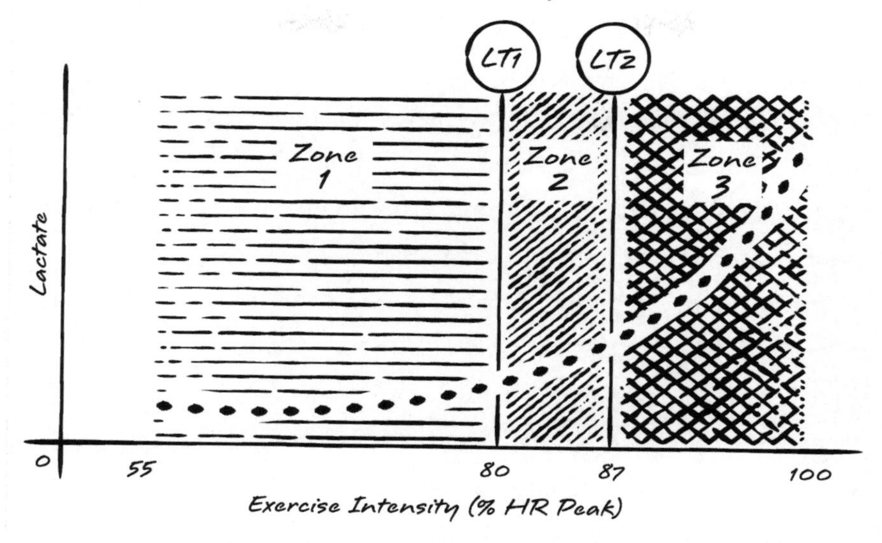

The Norwegian method relies on physiological markers—aerobic threshold (LT1) and anaerobic threshold (LT2)—which are defined by the lactate concentration in the athlete's bloodstream.

form—is that double-threshold sessions are often broken into intervals to control intensity to an even greater degree and allow time for testing.

How to define a "threshold" session will likely remain a debate in endurance sports literature until humans stop running, and every athlete will have very different thresholds—whether that threshold is marked by lactate, heart rate, perceived exhaustion, or some other metric. For the sake of simplicity and consistency, the two thresholds we will focus on are lactate threshold 1 (LT1) and lactate threshold 2 (LT2). These will vary wildly from athlete to athlete, but LT1 is essentially the intensity at which lactate begins to spike in the bloodstream, and LT2 spikes exponentially, requiring a reduction in intensity. Lactate isn't the actual culprit causing the slowdown, but it remains the most accurate indicator of when hard becomes too hard. LT2 is often referred to as anaerobic threshold, which will be more familiar to anyone with experience training for an explosive sport. Weight lifters and power-sport athletes are familiar with anaerobic threshold and surpass it multiple times per day. Endurance athletes—particularly stern devotees of the Norwegian method—are more likely to encounter it only in race-specific sessions.

If there is any term in the world of endurance that suffers from a lack of consensus more than *threshold*, it's *zone*. Some coaches swear there are only three zones, others prefer five, and many do away with zones altogether. The forefathers of the Norwegian method worked within a three-zone system that is essentially easy (zone 1), moderate (zone 2), and hard (zone 3). Ideally, the Norwegian method determines and monitors these zones via lactate. That's not to say it can't be

A Simplified Overview of Different Training Methodologies

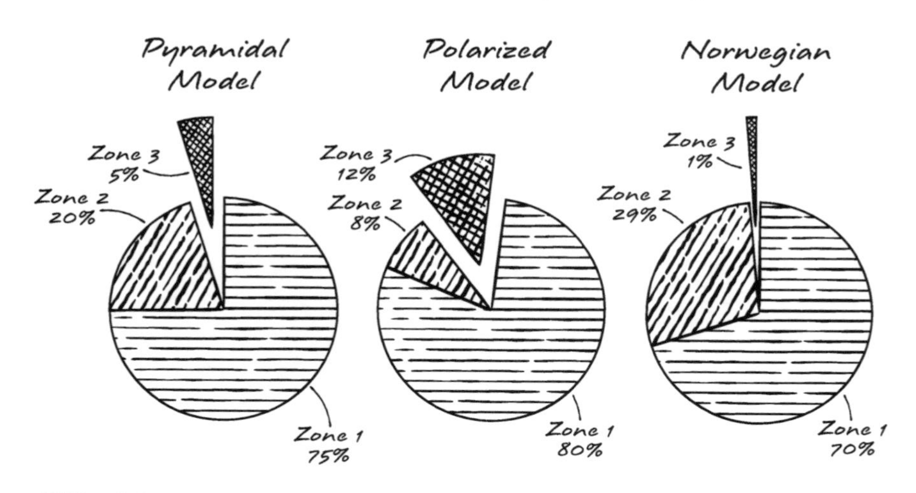

While all three models include plenty of volume at low intensity (zone 1), the Norwegian method typically puts in very little time above LT2.

replicated with heart rate, which has been traditionally used by pyramidal and polarized methodologies.

Let's consider this gross simplification as a starting point for differentiating the three models:

- **A pyramidal model** typically entails close to 75 percent of training in zone 1, 20 percent in zone 2, and the remaining 5 percent or so in zone 3.
- **A polarized model** also entails close to 80 percent in zone 1, 5–10 percent in zone 2, and the remaining 10–15 percent in zone 3.
- **A Norwegian model** is roughly 70 percent in zone 1, almost 30 percent in zone 2, and only 1–2 percent in zone 3.

A three-zone system does a good job of isolating the middle zone, which is the zone that's being controlled for with lactate testing, which will be explained further in Chapter 15. In that sense, it's perfect for the Norwegian model, and it works perfectly for most athletes training for events as long as a marathon or Ironman, whether they're ordinary or elite.

Endurance training with a five-zone system

Anyone who has trained for endurance events of varying distances knows that there are varying degrees of easy and hard training. Lumping them together into zone 1 or zone 3 doesn't necessarily work, which is why a five-zone system has been popularized around the globe, particularly among athletes training to race shorter and faster events. In reality, the Norwegian Olympic Committee and Olympiatoppen officially work with an eight-zone system, but both models are largely based on rowing and designed to cover all sports—from curling to powerlifting. Zones 6, 7, and 8 have no relevance for athletes competing in events lasting longer than a couple minutes, but let's consider how the five-zone system defines exertion, effectively divvying up zones 1 and 3:

- **Zone 1: Very easy.** This is often called recovery work. There's less lactate production occurring at this exertion, so well-trained athletes can essentially exercise all day in this zone, bearing any mechanical limitations.
- **Zone 2: Steady easy.** Here the body is beginning to produce a small amount of lactate, but it's still the case that this intensity can be maintained for many hours and nearly all day for elite athletes. The top end of this zone is LT1.
- **Zone 3: Tempo.** This is not quite race pace for most marathoners or long-distance runners. Heart rate and lactate spike, and athletes can usually sustain this intensity for a couple of hours.
- **Zone 4: Threshold.** The top end of this zone is LT2, where intensity must be reduced for the athlete to keep going. This intensity is only sustainable for 20–40 minutes.
- **Zone 5: VO_2 max.** This consists of speed and sprint work. For most athletes, this level of intensity is sustainable for less than two minutes.

Critics of the Norwegian method argue that lactate testing is overcomplicating a process that can simply be controlled via metrics like heart rate, but measuring lactate concentration does more reliably remove the gray area between zones. Unlike the polarized and pyramidal approaches, it doesn't waste time nitpicking between zones 1 and 2 or 3 and 4.

The Five-Zone Model for Exercise Intensity

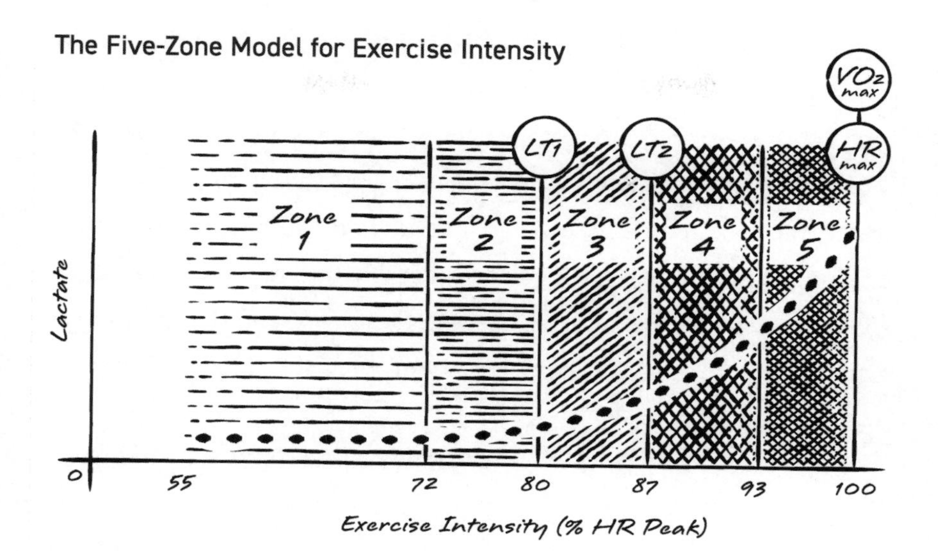

Endurance sports more commonly use a five-zone model. The physiological markers, LT1 and LT2, still define this model, but low-intensity (zones 1 and 2) and high-intensity (zones 4 and 5) work is further defined.

That doesn't mean you shouldn't, and especially for newer runners, five zones seem to be a pretty good starting place for learning about training intensities and, more importantly, how not to be stupid with them. A 2021 study of recreational triathletes training for an Olympic-distance triathlon and left largely to their own devices showed that athletes tend to drift into the so-called gray zone—what is commonly considered to be somewhere in the lower reaches of zone 3 on the five-zone scale—for a large part of their overall volume. That doesn't keep them from achieving their goal of finishing an Olympic-distance tri, but if the goal is to increase performance over a long period of time, spending too much time in the gray zone will probably limit how much both volume and intensity the athletes can put out.

The case against hurt-so-good training

It's not actually that recreational—or even chiseled—veterans of endurance are stupid. It's just that going hard feels good. It makes athletes feel like they accomplished something big in the session, and the endorphins that come with going a little hard can make the temptation to go even harder too much to resist. No one gets into endurance sports because they're looking for something easy, and the drive to do something hard that got them moving in the first place carries over into each workout.

It's certainly not specific to endurance either. Gym rats and powerlifters have lived by the "no pain, no gain" philosophy for the better part of a century, and many recreational weight lifters believe they're not going hard enough if they aren't sore for multiple days after hitting the gym. But even a sport as explosive as powerlifting—where the goal is a single rep at absolute maximum intensity—has learned that there's more cost than gain in going all-out and that higher frequency may be the ticket to bigger gains.

Once again, it was Norway leading the way in this new way of training—sort of. In 2000, the struggling Norwegian Powerlifting Team brought on a former West German weight lifter named Dietmer Wolf as its coach, who began what he called the Norwegian Frequency Project. The weight-lifting coach brought a unique approach to the much more technical sport of powerlifting (where lifts have to be performed with very precise movement patterns). The critical change was taking the powerlifters from lifting three times a week to six—something that was considered impossible in powerlifting circles. Maxing out on consecutive days was a surefire way to get injured. But Wolf's secret was that he didn't want athletes to max out until competition.

To prove the efficacy of his theory that six lighter days were better than three hard ones, he convinced the Norwegian School of Sports Science to conduct a study on his lifters, splitting them into two groups. One group continued to max out—or nearly max out—three days a week for 15 weeks, while the other performed their biggest lifts at 72–74 percent of max. (That number is somewhat arbitrary; the aim was for roughly three-quarters of max effort.) At the end of those 15 weeks, the eight powerlifters going six days a week had nearly a 10 percent average increase in max deadlift, squat, and bench press, while the group using the old, three-a-week method had an increase of only 5 percent. (It's worth noting that the greater average increase was due almost entirely to increases in squatting and benching. There was almost no difference in deadlift gains between the two groups.) A German coach using Norwegian sports science found that the sweet spot for weight training wasn't anywhere close to where most of the athletes were training—and it was keeping them from training enough to achieve maximum adaptation.

If there is magic to be found in the Norwegian method of endurance, it lies in that sweet spot where you can get aerobic benefits from hard efforts with little or no cost to future sessions. That's not necessarily unique to Norwegian training, but regular lactate testing works as a precision-guided missile to locate that sweet spot. By layering week after week of this protocol for a year or, better yet, years on end, you can raise your thresholds. This means that over time and consistent training, it's possible to produce less lactate at higher intensities and ride a harder line come race day.

For some athletes, that sweet spot seems to be smack-dab in the middle of zone 2 of the three-zone system or somewhere in the gray area between zones 3 and 4 of a five-zone system. This variability makes it easier to see why strict proponents of the Norwegian method have done away with zones altogether and instead let the lactate meter talk athletes out of being stupid.

Everything old is new again

When I signed up for my first Ironman at age 19, the first thing I did was pick up as many books as possible about Ironman training. At that time, everything I found was either written by Mark Allen or included dozens of references to his training. The six-time Ironman world champion was the first to truly master the Ironman and the sport of triathlon, and fortunately for us, he's spent a good part of the past 30 years writing about it. Allen's marathon times from the early '90s—late in his career—remained the best ever on the Big Island until the arrival of the Norwegians nearly 30 years later.

Zooming out, Allen's training wasn't vastly different from what the triathlon champions of today are doing—whether Norwegian or otherwise. He was routinely doing uncontrolled double-threshold days without having a fancy name for it. But he didn't have the same gadgets and tools at his disposal as the athletes of today—and even if he had, he probably wouldn't have used them very much. Allen was famous for taking nearly every workout to the point of exhaustion early in his career, an idea that was borrowed from his upbringing as a swimmer. (He didn't even start really running until age 25.) The "do more and do it faster" model has produced plenty of world-class swimmers, but it doesn't always translate to running, where the mechanical cost of going hard is exponentially greater.

In Dr. Tim Noakes's *The Lore of Running*, Allen is quoted as saying, "Do more faster really only worked for those so talented that their genetics were going to override the lunacy of their training and take them on to greatness anyway." This was a revelation that came relatively late in Allen's triathlon career, after meeting legendary running coach Phil Maffetone, who suggested that he slow down a bit, at least for part of the year. Allen's reasoning for burying himself at the end of workouts was more mental than physical. He was searching for something spiritual in those depths and frequently used shamanic retreats and sweat lodges to take his pain tolerance beyond the point of pain. Or something. This is where Allen's training became much more art (or religion) than science.

Regardless of what Allen was searching for in the depths of his training and racing, his trial and error at the sport's highest level provided a valuable road map for the athletes and coaches to come after him, whether that be for me coaching myself to my first Ironman finish in 12 hours and 12 minutes or Olav Aleksander Bu coaching Kristian Blummenfelt to the fastest Ironman performance of all time, which was about five hours faster. Allen's big breakthrough of 1989 came after that fortunate meeting with Maffetone, who convinced him to reduce the frequency of his intense sessions and ratchet up the volume of his easy ones. With Maffetone's advice, Allen unlocked the back half of the Ironman marathon and won his first of five straight (and six total) Kona titles in 1989. The "do more and do it faster" approach hadn't worked for an event as long as the Ironman, but "do more and exercise more control" put him on top of the podium in an event the world was just beginning to figure out.

Control is the name of the game

Elite athletes at the very height of their respective sport are fond of saying that they worked harder than others to get there. Some of the most talented athletes on earth like to say that they somehow overcame a lack of talent to become better than others blessed with more talent. Of course, this may be true in select instances, but the tip-top of any sport is occupied by humans who won the genetic lottery on some level.

And the pros are humans who have a hard time holding back. This approach might work for the young and naive or for sports with a relatively low risk of injury, like swimming. Many elite swimmers can go deep into the well multiple times per day for days on end. But when it comes to sports like running and triathlon—which are complicated by the effects of gravity, such as pounding the pavement—holding back can be the most important thing for a driven athlete to learn. And it typically takes a good coach or some really good tools to communicate this lesson in a convincing way. The history of both running and triathlon is littered with exceptional athletes whose careers were capped or cut short by burnout and injury. It's most often athletes who push themselves to the point of exhaustion as often as possible instead of only when it's absolutely necessary who find themselves dropped.

That doesn't mean the Norwegian method is somehow easier than models that promote slightly more volume at race intensity. One of the biggest misconceptions about it is that the hard training is somehow easier because it's often more controlled. The best Norwegian athletes practicing this method are all very public with their training on social media, and it's clear that the hard days are hard, especially as they approach racing.

The other common misconception about the Norwegian method—or the recent success of Norwegian athletes—is that the summer sports athletes started training like the Nordic skiers, and voilà, world records

A Different Take on the Norwegian Method

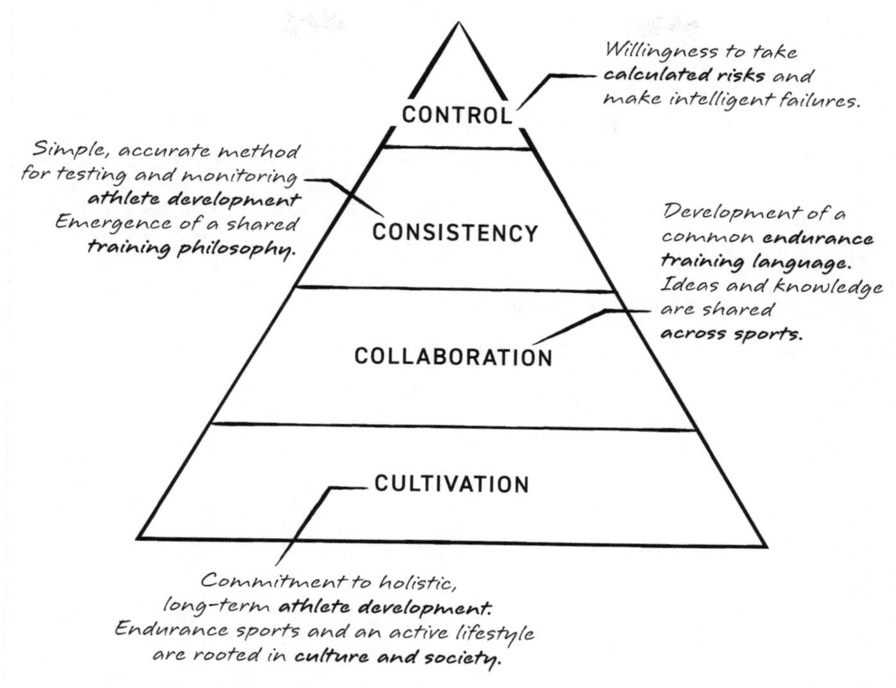

Dr. Stephen Seiler credits Norway's success in endurance sports to a biopsychosocial program that benefits from collaboration across sports, science, and practice.

started falling. The major similarity in training for the two sports is a lot of zone 1 volume. The predominant method of elite Nordic skiers is best described as polarized—a model that may not have started in Norway but was first conceptualized by an honorary Norwegian. Dr. Stephen Seiler has spent his life studying endurance athletes, mainly in his adopted home country. Much of his research has become the basis of our understanding of polarized training, and he believes the Norwegian method is just a new name for an old model. He frequently refers to the current iteration as the "Norwegian method 2.0," implying that polarized training was the original Norwegian way, and now there are just more accessories and testing. He might be right, but decades later, it's likely method 3.0 or 4.0.

The major distinctions between Seiler's conception of the polarized model and Marius Bakken's molding of the Norwegian method were the staunch dedication to lactate-guided sessions and defining the three zones via lactate concentration. As we'll learn in the next chapter, Marius took both the frequency and intensity control of lactate testing to extremes, essentially avoiding any and all zone 3 work. As is typically the case, it takes a zealot to incite a movement—or in this case, an athlete with a remarkable tolerance for trial, error, and blood.

HUMAN

7

MARIUS BAKKEN: THE GODFATHER

There is more honor in accumulating little by little
than in reaching for the sky and ending up flat on your face.

—VATNSDÆLA SAGA

At some point many decades ago, long before Norway was the envy of the endurance sports world, Nordic sports science was a tick above the world average, with its skiing, rowing, and running coaches pushing research faster than most. If you ask Marius Bakken who set the Norwegian train in motion all those years ago, he would say Leif Olav Alnes, who's still very much in the train's driver's seat today.

Alnes is rarely mentioned in the annals of endurance training because he's one of the world's greatest sprint coaches—currently guiding Norwegian 400-meter hurdle gold medalist Karsten Warholm—and endurance and sprinting have functioned like oil and vinegar. They only mix when they're forced to, but that happens quite often in Norway, and in this case, both disciplines have ended up better for it.

A lukewarm start-up for lactate testing

Norwegian runners began experimenting with lactate testing in the late 1990s, during the buildup to the 2000 Sydney Olympics. At the time, different groups of distance runners throughout Norway and Scandinavia were training with various approaches, like the popular Lydiard or Martin-Coe methods, both of which prescribed varying intensity at different phases of the training cycle and different phases of each workout. A lot of the best runners of the day, in Norway and elsewhere, were basically training all the intensities all the time.

When the Norwegian School of Sports Science, in partnership with the Norwegian Olympic Committee, began testing both summer and winter Olympians for blood lactate in controlled lab settings, it was hardly an initiative to create some radically new system of training. It was about collecting yet another data point to determine whether controlling the intensity of hard sessions in fact held merit versus doing nearly every hard session as close as possible to race speed—or even faster than race speed.

The project lost steam rather quickly, mostly because the prevailing wisdom of the day—a little more than a quarter century ago—was that race pace was something to strive for frequently in training; otherwise, there would be no way to sustain it come race day. That was the method prescribed by Peter Coe, a Brit who at the time coached a number of the top distance runners in the world, including his son, Sebastian. The younger Coe won Olympic gold in the 1500 in 1980 and 1984, which instantly made his dad the most in-demand coach on earth. One runner who came under Coe's tutelage was Norwegian track star Marius Bakken, who is considered by most to be the grandfather of the Norwegian method—even though he's barely old enough to be a grandfather himself.

Marius was one of the Norwegian runners who began experimenting with lactate-controlled threshold sessions in the late '90s, but unlike those involved with the Norwegian School of Sports Science, he never lost interest. That's because he was in the driver's seat of the experiment, and from that vantage point, he had the best view of just how well it was working when properly controlled. He just needed a few years and a few thousand pricks of his ears and fingers to find out what "properly controlled" actually meant. His belief—and the one shared by a small group of Norwegian runners who continued experimenting with lactate-guided threshold training—was that the main limiting factor for all distance runners was the anaerobic threshold. Increasing top-end speed is neat if you're training to race a lap or two around the track. Any longer and an athlete's ability to run at their anaerobic threshold for longer (or to run faster at that threshold) is much more important than being able to run well above it for a short time.

A tour of world-class coaching

Prior to Norway's most promising distance runner meeting its most legendary coach, Marius had to leave Norway to begin his journey to becoming the first Norwegian to make the Olympic finals of one of the marquee events on the track. Marius applied to spend his final year of high school in the United States through a unique exchange program between Norwegian and US students. A recruiter noticed that Marius mentioned that he was a good runner on his application, and Marius was offered the opportunity to spend his final year of high school, the 1995–1996 academic year, at York High School in Illinois training with legendary coach Joe Newton. Newton passed away in 2017, leaving York with 29 state titles and a smattering of national records throughout his 60-year career at the same place. On his blog, Marius credits Coach Newton for teaching him what hard work truly meant and for instilling a passion for running that he didn't have prior to York. "It was extremely hard training," he writes. "Probably the hardest year of training I ever had, and it showed me the road to the future. The closest you can get to Kenyan training—at 17—is at York. It gave me the running background in the future thanks to coach Newton's principles and enthusiasm."

The next few years would be a bit turbulent for the young runner as he bounced from place to place and coach to coach, learning a ton about different approaches to endurance running but ultimately taking his times on the track in the wrong direction. From Illinois, it was back to Norway for a year of virtual training via Peter Coe, whom he'd been introduced to by Coach Newton. Because Coe was the most sought-after coach in middle-distance running in 1997, if he agreed to be your coach, you weren't saying no. It was Coe who taught Marius all about periodization and about how hard intense training could be. He was routinely running interval workouts harder than race pace because Coe believed that the only way to run faster was to run faster, which Marius was beginning to find wasn't always the best way for him.

After a year of training with Coe, Marius was recruited to Indiana University by Sam Bell, who had just coached Bob Kennedy to become the first man to break 13 minutes in the 5000. Marius had once again hit the American coaching lottery. At 12:58, Kennedy's time was close to what Marius had in mind for himself, and now he had a coach who'd done it before.

The program that worked so well for Bell and Kennedy was very traditional: two hard workouts per week close to race intensity, one long run, and three or four moderate or recovery days. Marius knew he didn't have the same natural talent as Kennedy, so using the very same approach probably wasn't going to help him close the gap. He returned to Norway in 1998 and pondered what he could do better and kept coming back to the same thing: volume.

Whether he realized it or not at the time, the accumulation of knowledge from bouncing from world-class coach to world-class coach was about to pay off. Even if it didn't at Indiana. His progress stagnated—and even regressed a bit in the 1500—doing the exact same training that had worked so well for Bell, which, relative to what he had been doing at York, was low in volume. A second year in the American Midwest was enough, so it was back to Norway at age 20 with the idea to go long and try something new.

Marius took a trial-and-error approach to every aspect of his training—coaching included. He was always destined to be his own coach; he just needed to learn what aspects of each methodology could work for him and which ones needed to be discarded. Back in Norway and with the 2000 Olympics looming a little more than a year away, Marius solicited legendary Norwegian runner Per Halle to be his coach, and he wasn't asking for much. Basically, Marius asked Halle for an exact replica of his training plan from 1972, when he'd finished seventh at the Olympics in the 5000 in 13:34. Two years later, Marius ran a personal best of 13:27.

Marius had much bigger goals than 13:27, but he also had a tool in his pocket that his new coach did not. Halle was one of the first Nordic runners to log well over 100 miles week after week and obviously made huge concessions to intensity to handle that sort of training. Halle's prime year of 1972 was five years before a company called Polar invented the world's first wireless heart rate monitor in collaboration with the Finnish Nordic skiing team, so his only form of intensity control outside of a lab setting was by feel. Marius knew from his past three years and three coaches that more volume—not less—was what he responded to best. He knew he needed a lot of the "easy" stuff. He just needed to figure out what the right amount and degree of hard was.

After two years without logging a single week over 100 km (62 miles), Marius's first week with Halle's training plan had him running 180 km (over 110 miles), and he rarely came down from that number throughout the 1999 season, save for the occasional life hiccup or injury. He became obsessed with lactate testing and learning just how much threshold training his body could handle without jeopardizing the volume he knew he needed. He was laying the foundation for the Norwegian method throughout that year—or at least modernizing it using his takeaways from some of the best non-Norwegian coaches of the era, a training plan from 30 years earlier, and a fancy new blood tester that had him more zealous about training than ever before.

The next few seasons were the kind that every runner dreams of. Time fell off a handful of seconds at a time. He trained and raced longer, and nearly every time he competed over 5000 or 10,000 meters, he discovered a new PR. Legendary Swedish runner Lasse Viren's Nordic record of 13:16 had stood for 24 years until 21-year-old Marius shaved five seconds off it at a meet in Rome. That got him to his first Olympic Games in Sydney that same year, where he officially finished 25th after failing to advance out of the first heat.

Nonetheless, it showed that his method was working. So far his method mostly consisted of learning everything he could from a single coach over the course of one season, keeping what worked, ditching what didn't, and moving on to the next one. It wasn't that Marius was explicitly mercenary about his athletic development. Far from your typical coach-athlete relationship, Marius lined up a rotating cast of advisers, who were happy to offer feedback or suggestions to an eager and talented athlete while they observed his unconventional method unfold.

A Nordic runner in Africa

It wasn't just coaches he wanted to learn from. Of the 24 men who finished ahead of him at the Olympics, 12 were African, including the top six. At that time—and still to this day—most of the best distance runners on earth were living and training at altitude in East Africa or Morocco, and Marius knew he wasn't going to glean much from racing against them at the occasional meet. At the behest of his coach du jour, Frank Evertsen, he took three trips to Kenya during the 2001 season and added a fourth stay at altitude in Switzerland over the summer. That season ended with a new Norwegian record in the 5000 and a ninth-place finish at the world championships in Canada.

His evolving Norwegian method was becoming more Kenyan, and it was about more than just the altitude. It was about lifestyle and upbringing. The young man who grew up in a country that had ample resources wanted to know what Kenya was doing better when it came to building young runners. To find that answer, he sought out the godfather of Kenyan running, who happens to be an Irish missionary who came to Kenya in 1976 with absolutely no running or coaching background. In the long line of great endurance coaches to have zero experience with endurance growing up, Colm O'Connell was the first to have success at the very highest level. Colm came to Iten, a town in

Kenya's Great Rift Valley, in 1976 and planned to spend three months teaching geography and spreading the word of his Lord. He has lived there ever since and has helped produce some of the greatest long-distance runners the world has ever known. He has mostly worked with kids at the world-famous St. Patrick's High School, which has turned out a total of 25 world champions and four Olympic gold medalists since Colm showed up all those years ago.

Marius wanted St. Patrick's secret, and he suspected that there was no secret at all. The Kenyans were simply doing a lot more volume than he ever thought possible for kids as young as 14. The program at St. Patrick's entailed 13 run sessions every week, in addition to strength, stretching, and drill work. The regimen itself didn't surprise him very much. He knew that the biggest advantage the Kenyans had was that they were running a lot more from a much younger age—and at higher altitudes—than just about anywhere else on earth. What did surprise him was the amount of speed work everyone was doing and the fact that the programs of the best 800-meter runners were remarkably similar to those running 5000 meters or longer. In the 13 runs he observed them doing each week, he only labeled two as "easy." Two were what he would call VO_2 max sessions (i.e., very hard), and the other nine were executed close to the anaerobic threshold.

Marius obviously couldn't go back and retrain himself from age 14, and even if he could, he knew the Kenyan way of training young athletes wouldn't be successful for him and anyone who didn't grow up running in East Africa. As he posted on his blog after his first trip to Kenya in 2000 (he was as early to blogging as he was to lactate), "Try this on a 14-year-old Norwegian or American and let them enjoy it—at 2500 meters [i.e., at an elevation of 8,200 feet]!"

It was way more volume than he had previously thought possible for young athletes, and they were adapting to it at altitude—something Norwegian sports science was only just starting to appreciate. But his first few trips to Kenya had him thinking that he needed a lot more threshold work without sacrificing the 180 km of volume he knew he had to maintain. It also had him thinking that he needed a new coach and mentor, and perhaps one with more experience going faster than longer.

A meeting of the minds

He'd observed Alnes work with Norway's best sprinters during the buildup to the 2000 Olympics, and after witnessing what Colm's young group in Kenya was doing with mixing sprint training elements into an endurance program, Marius was more convinced than ever that perhaps the sprinters were doing certain things better. He was most interested in the workouts where Alnes's sprinters were doing six to eight hours of work in a single day, almost all at high intensity, with long periods of rest in between sessions. Oftentimes those eight hours of training were spread across four separate sessions. The idea of the double threshold was taking root; he just needed to fine-tune it alongside his newest adviser.

Marius and Alnes built a shared learning environment during training—the distance runner had a lot to learn from the sprint coach and vice versa. After a couple of seasons of stalled progress, the two began working closely together in the buildup to the 2004 Olympic Games in Athens. It's not that Alnes completely overhauled the system Marius was building; he just found a way to increase the overall load, which he believed was the limiting factor for 100-meter sprinters and marathoners alike. In the specific case of Marius, he believed adding more load required a makeover of his muscular system. He simply needed to get stronger before his skeletal and nervous system could handle the speedwork. He started doing real strength training—not just drills—for the first time. Alnes didn't send him to the gym to lift weights with the 100-meter runners. But he did add parachutes and resistance bands into his training regimen to target the specific adaptations required for running. In 2003, Marius ran his lifetime personal best of 13:06, qualifying for his second Olympics.

As is typically the case at the Olympics, the distance events are slow and tactical. His goal was always getting the most out of himself and finding the smartest way for him to continue to progress, and toward the end of his career, he had a level of satisfaction that couldn't be washed away from two lackluster runs on the sport's biggest stage. He wanted to end his career knowing that he had gotten everything possible out of his body, and his Norwegian record of 13:06 was close to what he believed his maximum ability to be. And it would stand until Jakob Ingebrigtsen lowered it all the way to 12:48 in 2021 and began his assault on the 5000m world record.

Yet another chance meeting

Marius began a gradual transition to coaching in the years that followed. He formed a small training group that spent the winters in South Africa, and it was there that he met a coach named Eric Toogood, who was the primary coach of Henrik Ingebrigtsen—the eldest of the three Ingebrigtsen brothers—before their father took over. Toogood was training a young steeplechaser named Bjønar Kristensen, who proved to be the perfect training partner for Marius, and the two athletes began doing nearly all their workouts together while in South Africa. Marius taught them to incorporate lactate and double thresholds twice a week, which Toogood then brought to Henrik, and the rest is history—depending on who is telling it. The Ingebrigtsen camp boasts that they might have taken over the world of distance running with a system of their own making, but as is always the case in coaching, they started by copying someone else's homework. They just happened to sidle up to a star pupil at the time. Had Marius never gone to South Africa to train with Kristensen, it's possible it would've taken much longer for the Ingebrigtsens to incorporate some aspects of his method.

Marius officially walked away from competition in 2006, and by then, he was already two years into medical school at the University of Oslo. Med school was always part of the plan, and some of it was relatively easy for the 27-year-old, who'd spent a decade learning everything possible about his blood and energy systems. He's now a family physician in his hometown of Kristiansand, outside of Oslo, and running is still very much a part of his life. He's just let himself ease off the science a bit. It's still the case that you may see him on the local track, drawing his own blood and chatting to Alnes about all things running—both fast and slow. The Norwegian method has become something of a lifestyle for its godfather, and even in his 40s, there's still plenty left to learn about his own body.

SCIENCE

8

MEXICO CITY AND THE BIRTH OF ALTITUDE TRAINING

Utenfor fjellet er det folk også.

Beyond the mountain, there are people too.

—NORWEGIAN PROVERB

The 1968 Mexico City Olympic Games were groundbreaking for two reasons. First, it was the introductory Olympics to Latin America, a region they've only returned to one time since (2016 Rio). As culturally significant as it was for the Games to be hosted in Mexico, the athletes competing were much more concerned with the second reason. At 7,349 feet (2,240 meters) above sea level, Mexico City is by far the highest city to host a Summer Olympiad, and it gave the world the first real case study of elite athletes exercising at altitude.

When the International Olympic Committee awarded the Games to Mexico City over Detroit, Buenos Aires, and Lyon, there was a bit of an uproar from federations at sea level that competing so high in the sky would be too difficult. Their concern was mostly a hypothesis because so few studies had been conducted on the effects of exercising at altitude, but there had been enough years of collective human existence to know that it gets harder to breathe the higher you go. The athletic federations were particularly concerned because the altitude would surely have more negative effects on the runners than anyone else.

They were both very wrong and very right. Every single men's sprint world record was shattered, plus the women's 100 and 200 meters. It was around 800 meters where things started to take a turn for the worse. The distance events (1500m and above) had the slowest winning times in decades. It was also the first time that the distance events were dominated by East Africans, who lived and trained at a very similar elevation to the one they experienced in Mexico City.

In his book *Summer of '68*, Tim Wendel discusses the significance of 1968 in the history of baseball in America and also notes that it was an inflection point in the way athletes from many sports trained: "After 1968, training at altitude, even if the events were going to be held at sea level, was seen to be a great way to prepare. What Mexico City did in an offhand way was point out that if you can prepare the body in the right way, you can exceed expectations in athletics."

What we understood about altitude prior to 1968

The Mexico City Games were hardly the first occasion for research on the effects of altitude on performance. Nobel Prize–winning physiologist A. V. Hill coined the term *oxygen debt* way back in 1925, referring mainly to how exercising at higher altitudes increased lactate and required more recovery time. By 1960, renowned researcher Bruno Balke had done extensive research on ordinary people to demonstrate that higher altitudes reduced oxygen supply by lowering the air pressure and diminishing the oxygen content of the blood. Balke was brought on as an adviser to the US Olympic Committee ahead of the '68 Games, and his advice was essentially the same as it would be today: Train at altitude as often as possible beforehand, and then get to Mexico City as early as possible before your event to acclimate.

The altitude of the 1968 Games became such a hot topic of conversation in the buildup to the event that, in 1966, the University of New Mexico hosted the International Symposium on the Effects of Altitude on Physical Performance, which was basically an opportunity for Balke to present his research to the International Olympic Committee and scientists from a number of different countries. He recommended a minimum of three weeks of altitude-specific training for every athlete in their final prep for the '68 Games and said that, realistically, six weeks are needed because the first week or two of altitude training will be spent acclimating.

The IOC's response was to do everything possible to bury Balke's research and claimed his research was incomplete. This was at the height of the IOC's amateurism battle, and their bylaws stated that no athlete could be away from their job or school for more than four weeks at a time; otherwise, they would be deemed professional. A six-week altitude camp was out of the question, and Dr. Balke was standing on a podium in New Mexico proclaiming that every athlete from every country needed to do exactly that to optimize performance.

Needless to say, only athletes and federations that were able to spend considerable time at altitude were willing and able to skirt the IOC's regulations, which certainly wasn't that hard to do in 1968. Representatives from Britain, France, the USSR, and both Germanys were on hand for Balke's talk and didn't care much for amateurism. East Germany was gearing up for its first Olympiad and would eventually become notorious for breaking every rule possible (along with a number of world records). The USSR built two high-altitude training centers in 1967 in Tsaghkadzor and Alma Ata, with the latter at the exact altitude of Mexico City. France sent most of its top summer athletes to train in Font Romeu for the very first time, which remains one of the most popular high-altitude destinations for endurance athletes today.

For the overwhelming majority of athletes, who didn't have the benefit of high-altitude living or camps, the Mexico City Games were harder than any competition they had experienced. In the first distance event on the track, the men's 10,000, Neftali Temu of Kenya won gold by less than a second over Ethiopian Mamo Wolde. The great Mohamed Gammoudi of Tunisia won bronze, followed by athletes from Mexico and the USSR. The entire top five lived at altitudes very close to that of Mexico City in the year leading up to the Games, but Temu's winning time of 29:27 was 1 minute and 48 seconds slower than the world record set by Australian Ron Clarke. The Aussie legend was the first of the non-altitude-trained runners to finish that 10,000, in sixth, and then collapsed to remain unconscious for the next 10 minutes. With only one event into the '68 Games, it was clear that they would be like none before—or after.

The ripple effects of those Games remain today. For the IOC, the big takeaway was that lower elevations are preferred by the vast majority. Every Summer Games since has taken place below 2,000 feet, with most at sea level. For many athletes and coaches, however, the big takeaway was that higher might be better—for training, at least.

A steep learning curve to climb

During the late '60s and early '70s, when the real world was dealing with the Vietnam and Cold Wars, the assassination of a US president, and disco, the world of sports science was in the midst of its biggest breakthrough in decades—especially for endurance athletes. High-altitude training (and living) became the focal point of sports science, and some researchers even began studying humans from extreme altitudes, like Bolivia and Ethiopia, to see what biological adaptations they'd undergone. The research from corners of the globe painted a very clear picture of something Balke was already very sure of: Long periods of time spent at altitudes above roughly 1,500 meters (4,921 feet) decrease oxygen hemoglobin saturation, which stimulates the kidneys to produce erythropoietin (EPO). Sports scientists, coaches, athletes, and charlatans across the globe were just beginning to understand the importance of EPO in endurance performance.

The research was so convincing that in 1978, the US Olympic Committee (USOC) moved its Olympic Training Center from sea level in New York City to Colorado Springs at 6,035 feet. By this point, they were well behind the rest of the world in moving their state-run training centers to higher elevations. Among those leading the charge on altitude research and training were the Swedes, whose Nordic skiers were well acquainted with the effects of living at sea level and racing at elevation. After winning a gold medal in the men's relay at the 1964 Innsbruck Olympics (7,369 feet above sea level), Swedish cross-country skier Karl-Åke Asph said in an interview that the effort had nearly killed him and that he should've gotten to Innsbruck even earlier than a few weeks out to acclimate. Sweden was one of the only teams that had bothered to show up to the Austrian Alps a month early. Asph said nothing about training at altitude prior to acclimating, which wouldn't become widespread until the 1980s.

The reason why athletes and researchers focused on altitude acclimatization and not training for so long was that the benefit was clear and quantifiable. Trying to quantify the benefit of, say, a four-week altitude camp on an event a month or two later—perhaps at sea level—is more complex, and it's something sports scientists are still struggling to master today. It doesn't help that every athlete responds differently to altitude, with some having a pronounced response and others barely registering a noticeable response. In a controlled setting, it's hard to debate that altitude training has an effect on athletes: It does increase the amount of oxygen-rich red blood cells circulating in their system, and, in theory, that should increase performance. In theory. In practice, it's a bit more complicated, and it's something Norwegian athletes and coaches have been experimenting with for at least the past 40 years.

A Longitudinal Study of Norwegian Rowing's Comeback

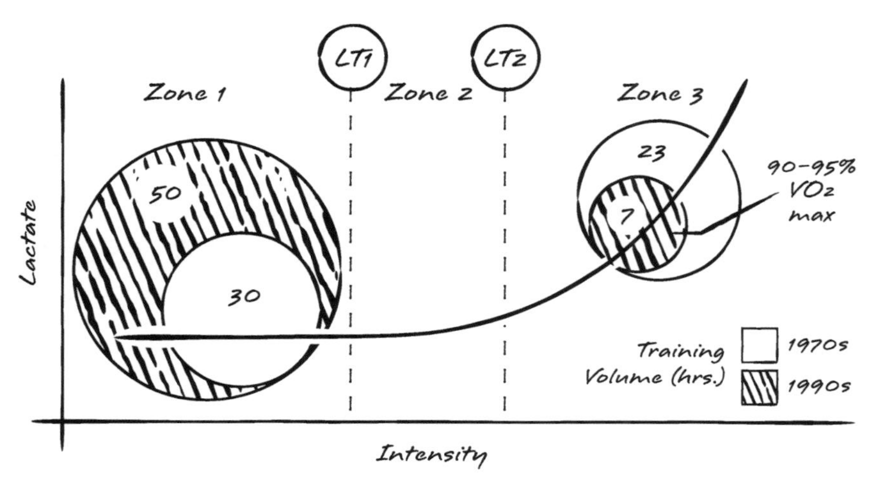

Between 1970 and 2001, elite Norwegian rowers increased easy training from 30 hours per week to 50, and decreased VO_2 max training from 23 hours to just 7. With the introduction of altitude training, they saw an average of 10% boost in VO_2 max.

Training camps in the sky

A longitudinal study published in the *Scandinavian Journal of Medicine and Science in Sports* in 2004 looked at changes in 28 Norwegian rowers between 1970 and 2001. This span of time saw a transformation in Norwegian sports, as we've previously seen. Rowing was one of the disciplines Norway had excelled at in the early 20th century, and they were looking to rekindle that greatness. The study was coauthored by Dr. Seiler, and it would ultimately become foundational to his concept of polarized training and further development of Norwegian training models. It remains one of the best longitudinal studies we have on endurance athletes because the Norwegian rowers tested and tracked a lot of metrics. Unfortunately, it's specific to athletes competing in roughly six-minute intervals and events. Between 1970 and 1990, Norwegian rowers, on average, increased their VO_2 max by 12 percent and improved their six-minute power by 10 percent. The authors identified three primary changes during this time that accounted for these rather seismic leaps:

1. Low-lactate training (less than 2 mmol/L) increased from 30 hours per month to an average of 50.
2. Race-pace (suprathreshold) training decreased from 23 hours to a "mere" 7 hours.
3. Altitude training was introduced for the first time.

The 1970–2001 time frame was chosen for the study because it traced the acceleration of Norway and Olympiatoppen's application of sports science from rudimentary to world-class. Norwegian rowers began experimenting with altitude training in the late 1970s, and during the '80s, it became a staple of their peak phase, which usually happened twice per year. The program was designed by Rolf Sæterdal, the rowing coach and physiologist who Arild introduced us to earlier. A few weeks prior to the most significant competition, they would spend two or three weeks at roughly 7,000 feet, returning to sea level with just enough time to recover before the regatta (typically another two or three weeks).

In the 1990s, however, there was a shift in the Norwegian approach to altitude, moving the stimulus from precompetition peaking to offseason priming. The idea was to boost an athlete's basic conditioning with multiple altitude camps during the offseason to begin each year at a better baseline. (Who wouldn't want a natural boost of EPO at the start of a new season?) In the 1980s, the Norwegian National Rowing Team did 26 precompetition altitude camps and just two camps in the offseason. In the 1990s, the breakdown pivoted to 14 precompetition camps and 36 winter training camps throughout those 10 years.

According to Espen Tønnessen, a sports scientist who works with Olympiatoppen, it was the Norwegian Rowing Federation that led the charge in altitude training, with the cross-country skiers next in line and finally the runners in the late '90s. It was the federations leading the way and sharing information. Olympiatoppen didn't issue altitude guidelines for all federations until the build-up to the 2002 Winter Olympics in Salt Lake City.

Dialing it in

In making the final preparations for the Salt Lake Games and the following Summer Olympics in Athens, Ørjan Madsen and Jim Stray-Gunderson, who had done extensive altitude research with German athletes, were specifically hired by Olympiatoppen to focus on the altitude training of the endurance athletes. Madsen is one of the only great swimmers in Norway's history and is partly responsible for the beautiful natatorium in Bergen, where he can still be found coaching. While Kristian may be able to walk around Bergen without being noticed, Madsen can hardly make it a block in his lifelong hometown without running into an athlete he's mentored who is eager to catch up. Between 1997 and 2005, using both their own research and just about every study they could get their hands on going back to 1968, the duo of Madsen and Stray-Gunderson designed a philosophy of altitude training that still acts as a blueprint for today's athletes from Norway and beyond.

At least two altitude camps per year were planned in accordance with the competition schedule, with more being ideal if it worked with the season's plan and the athletes' families. The goal was to do a high volume of low-intensity training during a three- to four-week stay at altitude and then reduce training time and prioritize training in the highest zones (often above threshold or LT2) once an athlete returned from altitude. The expected results were an increase in VO_2 max and preparation for the specific demands of the competition.

Norway's cities are all at sea level, and living a normal life at high altitude isn't really an option, so Madsen and Stray-Gunderson's research mainly involved athletes at month-long altitude camps versus living at altitude. The researchers were keenly aware that higher altitude was usually better—as long as an athlete had time to recover and prepare for competition altitude—so much so that they pushed for altitude chambers at the Olympiatoppen headquarters in Oslo and altitude tents for as many top athletes as the Norwegian Olympic Committee would fund.

They did, however, point out a unique benefit of altitude camps that living at altitude doesn't afford—it's what they and others call the *gathering effect*. It's something skeptics of altitude training (yes, there are a few) have called out in recent years, and it's hard to argue with its existence. Going away from home for a few weeks at a time to train in a beautiful place with like-minded people is generally a good way to improve fitness—whether it's at altitude or not. A month-long focus on training and recovery will boost fitness at sea level or at 7,000 feet, but by the early 2000s, it was clear that most athletes would receive the greatest boost if those camps were done at higher elevations. Better to have that gathering effect happen up high.

After nearly a decade of research on both Norwegian athletes and others (Stray-Gunderson, in particular, did extensive research on American runners and worked with a number of prominent coaches), the duo laid out the following guidelines that are still used by many Olympiatoppen athletes—particularly rowers—today. (It's important to note that most of their research and guidelines were focused on rowing, which often uses an eight-zone system. When they refer to zones 4–8, it's the upper end of zone 3 for a three-zone system or zones 4 and 5 for a five-zone one.)

» Athletes should live between 2,000 and 2,500 meters (6,550–8,200 feet) while at altitude camp for at least 16 hours per day.
» Duration of camp must be a minimum of three weeks, but the hemoglobin effect increases with duration of stay.
» Low-intensity training can be done at the same altitude as living altitude.
» High-intensity training (zones 4–7) should be done slightly lower than living altitude.
» The camp should be bookended with training in zones 1–2 only. This includes the two to four days before the camp, the first two

days at altitude, the last two days at altitude, and the first two to four days back at normal elevation.

- Intensity control is necessary to increase total training volume.
- One zone 8 session per week is necessary to maintain strength/speed/VO_2 max.
- Aim for three camps and at least 60 days per year at altitude to improve performance at low altitude.
- Perform lactate testing a week before and a week after each altitude camp.

It's impossible to quantify just how much of the improvement Norwegian rowers saw in their VO_2 max and power outputs during the '80s and '90s was due to the introduction of altitude and how much was due to the other components of an increasingly dialed method of polarized training (which no one was calling it at the time). But the success of a system that prioritized altitude was unarguable.

If the goal of each Norwegian Olympic Federation is to produce a single athlete that is world-best caliber—which is something all of them except Nordic skiing emphasize—the Norwegian rowers nailed it with their development process. When your country is the size of a large city, hoping for more than one athlete with gold medal potential in each sport is a reach. What they've recently achieved in triathlon is an anomaly and mostly the result of two exceptional athletes pushing each other to new heights. On the heels of 20 years of some of the most advanced training on earth, the Norwegian Rowing Federation produced Olaf Tufte, who dominated single-scull rowing (the hardest kind) in the 2000s in a way the rowing world hadn't seen in a very long time. He competed in seven Olympic Games and medaled in four, including winning gold in 2004 and 2008. His coach was Tore Ovrebo, who we previously met as the boastful head of Olympiatoppen, a job he took over after overseeing Tufte's success, much of it at altitude.

Live-high, train-low protocols for endurance sports

At the very same time Ovrebo was orchestrating a training system to restore Norwegian rowing to greatness, Marius was designing his Norwegian method for running, and altitude became an increasingly important layer of the cake. He did a number of camps in Kenya, South Africa, and Bolivia (all of which offer access to extreme altitude) to both test and learn from those running cultures and to maximize his own aerobic capacity. According to Marius, lactate testing became even more critical during altitude camps because it was much easier to inadvertently overload the system at altitude, so intensity control was more important than ever. In his experience training in Kenya, intensity control was something that most of those runners knew inherently—without lactate testing. When he arrived in the Rift Valley with his portable lactate meter, it was the first time many of Kenya's best distance runners had their blood drawn and tested. Marius found that they were doing their "threshold" sessions even lower than 3.0 mmol/L.

The running cultures that Marius was training with and studying late in his career had a distinct advantage that he and his compatriots did not: They spent all their time in thin air and had been incurring all the hemoglobin benefits that come with it since even before they were born. In the late '90s, Stray-Gunderson's research shifted to the concept of "live high, train low" (LHTL), something he and fellow altitude researcher Ben Levine conceptualized in a 1997 study. The popularity of LHTL skyrocketed in the early 2000s, particularly with American

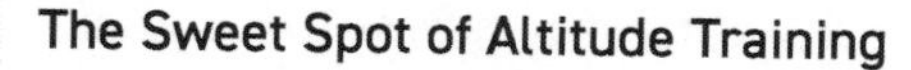

The Sweet Spot of Altitude Training

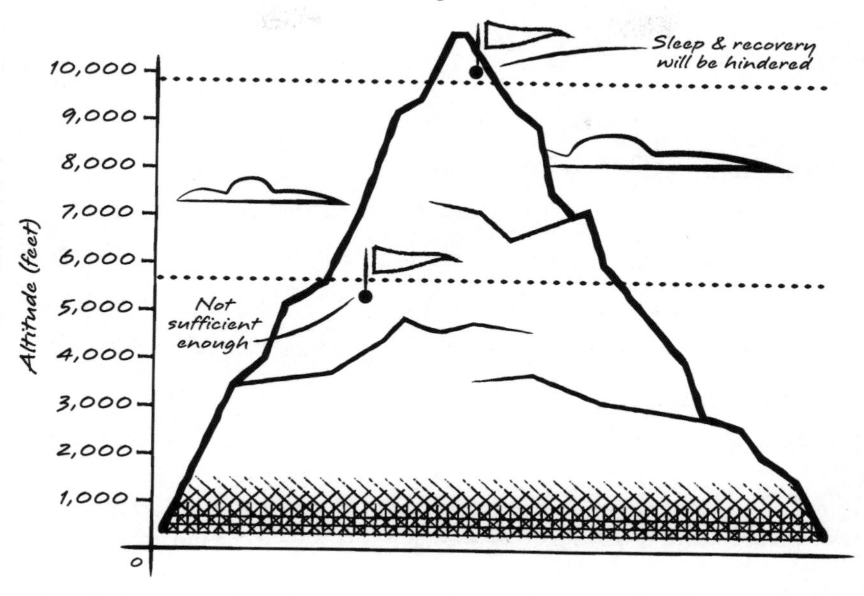

Athletes need to spend a minimum of three weeks training at altitude, reducing volume in week 1. The performance effects are thought to extend for three weeks.

runners, who flocked to places like Flagstaff, Arizona, where Norwegian rowers and skiers had been attending altitude camps since the 1980s.

The LHTL model wasn't drastically different from the altitude guidelines of Olympiatoppen (which were also based on Stray-Gunderson's research), but its popularization had athletes and coaches experimenting with altitude en masse and provided a pretty nice dataset of athletes training at altitude over the past quarter century. After Stray-Gunderson passed away in 2022, Levine, alongside University of Western Australia altitude scientist Oliver Girard, honored him with a new study in the *International Journal of Sports Physiology*, which looked back on the effects of 25 years of LHTL. After studying a number of groups who had used variations of the LHTL philosophy, the authors identified a few important takeaways of what we know for sure about altitude training:

- » **Elevation matters.** Below 5,900 feet (1,800 meters) isn't sufficient enough to produce measurable adaptation. If an athlete goes over 9,800 feet (3,000 meters), sleep and recovery will be hindered.
- » **Iron is key.** The authors recommend ferritin stores of 20 micrograms per liter for women and 30 for men. Even athletes with plenty of iron in their blood should consider supplementing while living at altitude.
- » **Three weeks is the minimum.** Any shorter is not *living* high. Longer is better.
- » **In week 1,** at altitude, an athlete should reduce training volume by 25 percent.
- » **Performance gains.** The authors noted that most athletes race well during the first week they come down from altitude. The second week is typically less favorable because the body has readjusted to sea level. However, they noted that the following three weeks after being at altitude can be another good window, which reflects the standard practice to allow for proper recovery when traveling to races.

The authors also noted that there was no difference between real and simulated altitudes. Altitude tents and chambers have gotten to the point that they can truly re-create mountain air at sea level, but the catch is that athletes have to be willing to spend at least 12 hours a day in there. That's a bit too intrusive on one's lifestyle for most people.

They also noted that athlete reactions to altitude are highly individualized. It's something Olav has likened to heat training and adaptation. What works for one athlete may not work for another, and, like most aspects of the Norwegian method, it takes a lot of trial and error to tailor altitude training to the individual. Olav has experimented with bringing his athletes down from altitude with as little as three days or as many as four weeks prior to race day and has found success at various places. Prior to the 2020 Tokyo Olympics (which took place in 2021), Kristian left altitude 12 days before the triathlon, which calls into question the widely adopted philosophy that the second week is a bad week to race.

Clearly, there's still a lot to learn, and Norway hopes to lead the way—much as it did at the inception of altitude research in the buildup and aftermath of the 1968 Mexico City Olympic Games. Without altitude driving a resurgence in Norwegian sports science in the '60s and '70s, the Norwegian method as we know it today would be something very different and probably wouldn't be an increasingly familiar term in endurance cliques these days. It's also what continues to be a catalyst for the continued evolution of the method today, along with heat, which we'll learn more about in Chapter 13.

HUMAN

9

OLAV ALEKSANDER BU: THE MASTERMIND

Wise he is deemed who can question well, and also answer back:
the sons of men can no secret make of the tidings told in their midst.

—HÁVAMÁL

We all have moments in time that change the trajectory of our lives: getting laid off from a job, experiencing the birth of a child, mourning the death of someone close. Few have experienced a moment as intense as Olav Aleksander Bu, and it was one that drastically changed the course of his life at age 30.

Olav grew up on a modest dairy farm in Eidfjord, situated along one of Norway's most impressive fjords that cuts into the center of the country. Many triathletes know it as the home of the Norseman Xtreme Triathlon, which has unofficially held the title of "world's toughest tri" since it debuted in 2003. The race finishes 3,200 feet above the fjord where it starts, to give you an idea of the terrain Olav grew up around. It was a quiet upbringing that involved a lot of hard work nearly all hours of the day. While he may have resented it a bit at the time, it instilled in him valuable lessons that have made him one of the most successful endurance coaches of this century. It's particularly impressive because he didn't start participating in triathlons until that fateful incident and didn't start coaching for a few more years.

There were no triathlons in Eidfjord growing up, and endurance sports were mostly the last thing on Olav's mind. Of course he skied a bit—everyone in Eidfjord does—but his athletic interest gravitated toward one of Norway's other great pastimes: sailing. It wasn't so much that he had an extraordinary love of the water or that he came from a long line of sailors, as is often the case. He just loved the complexity of racing sailboats, from the boats themselves, to the sails, to the timing of the tacks, to the way different teams of people try to manipulate the same force of nature better than others. He'd known he wanted to be an engineer from a young age—largely to eventually get out of Eidfjord—and sailing was the perfect athletic outlet for an aspiring engineer who spent the little free time he had dismantling every piece of technology his parents brought to the farm and putting it back together, usually successfully. What started with small things like a remote control at age 6 turned into larger and more complex devices, like a TV, by age 12. Nothing was off limits.

College entailed Olav procuring an electronics engineering degree in Oslo, as he'd always planned, and the years that followed went on rather smoothly and happily in the country that makes it easy to do just that. It's especially easy when you've always been the person at the top of your class. By adulthood, sailing became more of a hobby for Olav than an elite pursuit. Still, the idea of working for someone else's business didn't make any sense to him, so instead he founded two start-ups of his own, both of which were anchored in his favorite hobby and one that is still going very strong today, albeit without his leadership. Kitemill is a revolution in wind turbine technology that was born out of Olav's curious mind as he watched the sails harness wind and turn it into energy—in the form of velocity—on a boat. He's a big believer in energy diversity—another thing Norway does just about better than any place on earth—and wind power was of particular interest because he'd spent so much of his life thinking about capturing wind. He explains that, on a traditional wind turbine, the 10 percent of the massive rotor blades that constitute the tip harnesses 90 percent of the energy. The rest of the blade, and the tower—which extends deep and wide into the earth—is there to support the incredible force being applied on the tip of the blades, which is harnessed by a gearbox and generator in the hub around which it rotates.

If you've ever driven past a wind turbine—which every person in the American Midwest has at this point—you realize what an enormous structure it is, and it's all there to support the tips of the blades. Olav's idea in founding Kitemill was that perhaps there was a way to use kites, or small sails, attached to a more flexible support mechanism rooted in the ground. In theory, they can remove the support tower altogether and basically fly a kite (which looks more like a drone) in circles to capture the same amount of energy as a rotor-blade tip without all the extra structure. It's a brilliant concept that Kitemill and a few other companies with some of the richest investors on earth hope to have off the ground—literally—by 2030. That was too long for Olav to wait, and he was ready to build something new, so just before 30, he sold off his second company and was doing very well, even by Norwegian standards. It was onto engineering a new green energy idea—or so he thought.

A crucible for change

Everything changed on the weekend that was supposed to be Olav's 30th birthday celebration. Thanks in some part to his success in the business world, his family was able to realize a cross-generational dream of a dairy farm retreat on Norway's northwest coast in what is one of the most remote places in all of Europe. So remote that the only way to get people and supplies to the site initially was via a day-long hike or a short helicopter ride from the nearest village across multiple fjords and mountain passes. A road had yet to be constructed. The family's first getaway to their new retreat involved two loads of people and supplies via helicopter. Olav was on the first trip with his parents and a few others, but a mechanical issue sent the second helicopter containing his sister and three siblings-in-law cascading into the mountainside, killing all aboard instantly. It had happened so close to the destination that Olav and his family witnessed the horrific crash, but their ability to respond and alert the authorities proved a unique challenge. Part of the appeal of such a remote hideout was to be as removed from civilization as possible. That meant cellular reception was only available at the highest point above their home, some 3 miles and 1,000 feet above where they stood.

Olav had never done a long-distance run of any kind in his life, but he was the fittest of the group and was on his way up the mountain within minutes of the tragedy. His immediate response—before tears or anger or panic—was to take the next logical step. To control his emotions and do what had to be done in the moment. That moment—and his response to it in the immediate and distant aftermath—is something he still comes back to today, even if he's not fond of speaking about it. Who would be? He's repeatedly pointed to it as the most pivotal moment of his life up to this point, and in his mind, there's nothing more to discuss.

The reason others are often so interested in his tragedy is because the way he responded was so uncommon. When faced with tragedies, some people are stricken with post-traumatic stress disorder (PTSD) and depression—and indeed, some of his family were and never completely recovered—while others, like Olav, have a compelling need to turn that negative into some sort of positive. While he was running up the mountain to call the authorities, as well as his wife, who hadn't made the trip, he was more driven than ever before. In that struggle up the mountain to make the two hardest calls he'd ever made, he came to an epiphany. The man who had never really wasted a minute of his life was more determined than ever to never waste another minute. He was more determined than ever to become the best version of himself.

He went looking for an endurance challenge in 2011 and naturally came to the one in the town where he grew up. Why not make Norseman, one of the hardest endurance races on earth, your first? Part of the appeal was in the uniqueness of the challenge. He brought to triathlon the mindset of a sailor: Do something as hard as you can for a few seconds with the hope of maintaining speed for as long as possible before doing it again. Now in this new challenge, he needed to learn how to do the same thing over and over—for hours on end—while sustaining speed and effort as long as possible.

Perhaps because he chose one of the most grueling endurance events on earth as his first, Olav became absorbed in the challenge of racing for hours on end. He had managed to finish the race even though he'd had no idea how to properly train for it, and he suspected that even those who finished hours ahead of him weren't doing things in the best possible manner. It wasn't so much that he wanted to spur his own triathlon success; he wanted to learn what the best were doing and if it could be done better. He wanted to take human physiology apart and see if it could be put back together in a way to improve human performance—particularly for very long distances. In his new endeavor to engineer physiology rather than electronics, or kites, he did the same thing he had done when he started his first business: He went looking for a mentor. His search for Norway's foremost expert in exercise physiology led him first to the Norwegian School of Sports Science in Oslo, but the more he pressed about endurance, the more people told him that he needed to talk to Dr. Ørjan Madsen, who was probably at the fancy natatorium on the other side of the country.

Coaching worlds converge in Bergen

By fate or circumstance, the impressive natatorium in Bergen has turned into something of an epicenter of the Norwegian method, at least as it pertains to triathlon. It's where Kristian first met Roger, Arild first met Kristian and Gustav, Ørjan first met Olav, Olav first met Arild, and then Olav first met Kristian and Gustav. Without Ørjan Madsen, there would not be a beautiful natatorium in Bergen. And without that pool, the rise of Norway's triathlon world champions might never have happened. Madsen was the lone swimmer to represent Norway at the high-altitude 1968 Olympics and, in the years that followed, became the coach of the Norwegian National Swim Team and an obsessive researcher of exercise physiology. One Norwegian swimmer with an Olympic ticket set it all in motion.

In the years after the 1968 Games, Madsen became the head coach of the Norwegian National Swim Team, more or less got the natatorium in Bergen built, and turned Norway from perhaps the worst swimming nation in Europe to being almost competitive with their Swedish neighbors. At least for a minute. Madsen's commitment to making swimming relevant in Norway produced Alexander Dale Oen, who remains the only true world-class swimmer to come out of Norway. He's the reason a young Kristian wanted to pursue swimming instead of a more common Nordic sport. (Or his newest pursuit of cycling.) Alexander finished seventh at the 2005 World Aquatics Championship in the 100-meter breaststroke, which was a remarkable achievement for a Norwegian swimmer. Tragically, Oen died of a heart attack from an undiagnosed chronic heart disease at just 27, and Madsen's dream of seeing a Norwegian swimmer on an Olympic podium was mostly gone with him.

Today, the pool that was built on the heels of the success of the athlete Madsen had coached is finally home to an Olympic gold medal, it just belongs to Kristian, the swimmer-turned-triathlete with big goals who showed up shortly before Oen's death. So when Madsen received a random phone call from Olav saying that he wanted to learn everything possible about applying scientific training to his triathlon regime, Madsen suggested that he come to Bergen to meet the young triathlete and his new coach, who were attempting to do just that.

An outsider sees it differently

Olav became something of an understudy for both Arild, who had become the coach of the new triathlon team, and Madsen, even though his goals, initially, were quite different. Ever the entrepreneur, Olav sought out new technologies that he thought could help triathletes train smarter and new ways to collect and analyze all the data that should be available from all the hours of training. He was fascinated by lactate testing, the data it could provide, and how it could bring focus and purpose to each workout.

By 2016, Olav was brought on as an adviser to the national team and was asked to shadow the very small Norwegian triathlon contingent at the Rio Olympics, which was essentially just Kristian and Arild. A small part of a childhood dream of making it to the Olympics was realized, but as he watched the triathlons unfold, he wasn't all that impressed. He was impressed with three athletes: Alistair Brownlee, Jonathan Brownlee, and Gwen Jorgensen, who were in a different zip code from their competition on the run. The fact that the rest of the best athletes in the world could be so far behind had Olav thinking there was still a lot to be exploited in triathlon training. He didn't see any reason why Kristian—who had finished two and a half minutes after the Brownlees—couldn't be the one running away from the field four (or five) years later in Tokyo.

Over the course of the Olympic cycle that followed—the longest ever due to the pandemic pushing Tokyo 2020 to 2021—Olav went from an aspiring sports scientist to the coach who was making humans swim, bike, and run faster than ever before. Olav eventually took over primary coaching duties for Kristian and Gustav, and the trio took over the sport over the course of the next six years. The "boys," as Olav formerly called them, won six world titles during those six years to go along with Kristian's Olympic gold. They were winning short- and long-distance events all at once, something previously assumed impossible.

Oftentimes, it takes an outsider to move a sport forward. It's not uncommon for the best coaches to have little or no experience in the sport they're training athletes for. In running, some of the greatest coaches, like Joe Vigil, Colm O'Connell, and Jack Daniels, had never run themselves and never would. Especially in a sport as young as triathlon, Olav thought there may be a big chunk of time that could be lobbed off by being a little smarter. And by committing to a lifestyle that probably seems a bit deranged to most. Just as Arild seeing Marius use lactate testing planted the seed that there was another level of intensity control, Olav seeing Arild do the same made him wonder what other data and biometrics they could be mining. And he had two athletes who were ready and willing. Olav went about setting up the microculture that he believes has enabled them to figure out the sport in a better way than any men before them. As much as others want to learn from their techniques and methodologies, the one thing that can't be replicated is the dynamic of three distinct personalities that work so well together.

Exclusivity minimizes the variables

Not to mention a coach who has committed himself to exactly two athletes and no more. There is occasionally a small cast of training partners who come in and out of their training camps via exclusive invite. These are very often athletes from the Japanese National Team, which has formed something of an unofficial partnership with the Olav-Kristian-Gustav triumvirate, since the trio spent so much time preparing there ahead of the 2021 Olympics. The Norwegians wanted to train for the specific heat and humidity demands of Tokyo in July, and the Japanese were eager to learn whatever they could from the athletes and coach at the top of the sport. The Japanese have one of the older triathlon federations on earth, started in 1994, but they haven't had much success at the highest ranks. Their best result was fifth place at the 2008 Beijing Games by Juri Ide, who has become something of an endurance legend in run-crazed Japan for her longevity at the elite level, still going into her 40s.

In the years since the Tokyo Games and Kristian's gold medal, Olav, Kristian, and Gustav have broken ties with the Norwegian National Team, which isn't all that uncommon after a massive success. Every party wants to take credit for their part in achieving the one thing they set out to do, and certainly plenty of people had a hand in getting a boy from Bergen to the top of the Olympic podium in a sport Norway had no prior history in.

If you ask the boy from Bergen who deserves the most credit—other than himself, of course—it's Olav, and it's not a question he has to think much about. Without him, he no doubt had the talent and coaching to be one of the best in the world. He finished 13th at his first Olympics while they were just getting to know each other. But 13th to 1st is a massive performance gap and one he closed only when Olav took over the program and the three men set off mostly on their own. Olav has been willing to share a significant portion of the science that informs their training because he knows the real advantage lies elsewhere.

As he puts it, "I would not be able to make anybody the best athlete in the world because it comes down to chemistry—an environment where we are extracting the most from each other. With experience, you can get really far, but that last part requires something more than science."

CULTURE

10

MODERN NORWAY AND THE GOLDEN GENERATION

Berre den som vandrer finn nye vegar.
Only he who wanders finds new ways.

—NORWEGIAN PROVERB

Outside of the world of sport, Norway is best known for doing a lot of things better than everyone else. If there's a list published about quality of life, Norway is always at or near the top. Their only competition for any sort of social index rankings is typically their Scandinavian neighbors, with Denmark frequently laying claim to the happiest people on earth, likely on account of having slightly better weather than Norway.

By any measure other than weather, life is pretty good in Norway. It's not hard to understand why. It's a country roughly the same size as Arizona, with roughly the same population as the Phoenix metro area. There aren't a lot of people in Norway, they have tremendous resources, the population isn't growing rapidly (the average household has 2.11 children), and they've been very sensible with their politics and money over the past century—which makes them a bit of an anomaly on earth.

It's a country where socialism works, and if you talk to any Norwegian, there's a roughly 80 percent chance that they're happy and proud of their government and social structure. That kind of contentment with the government is reserved for a select few nations on earth.

After winning his first Ironman 70.3 world title, Gustav Iden was asked about the best part of growing up as an athlete in Norway. He went on to credit his homeland for providing a safe space to explore the edge of possibility. In his words, young athletes in Norway aren't afraid to fail, because if they do, they know they'll be OK. There will always be a decent-paying job of some kind waiting for them, even if it's not in an elite sport. There's a social safety net for athletes—and everyone else—in Norway that exists in only a handful of countries worldwide. Imagine what goals you might attempt if the fallout of failure was more contained.

If a sturdy social safety net were the biggest factor in cultivating world-class athletes, the Arab Gulf nations and Singapore would be a lot better at sports. Raising sturdy kids with the foundation needed to consistently become the best in the world in activities as physically and mentally challenging as running, triathlon, cycling, or skiing takes a village—or, in this case, a country. A 2020 study published in *Lancet*, England's leading children's health journal, ranked children between the ages of 0 and 18 from nearly every country in the world by a handful of mental and physical metrics. These included very basic things like nutrition and access to health care to more abstract criteria like life satisfaction and general happiness. To the surprise of very few, Norway came out on top, with their Scandinavian neighbors (and Finland) close behind. The United States finished 39th.

A major takeaway from the study—and subsequent studies in *Lancet* about youth and adolescent culture in Scandinavia—is that access and impetus to play outdoors as often as possible are what make the pagans of the north perhaps a bit better at raising children than the rest of us. It also highlighted the level of free play Norwegian children have relative to other Western countries and the sentiment that kids should learn to take ownership of their actions as soon as they begin to act. That's something that bleeds heavily into the Norwegian model of sport. Ownership and skills come first, and true competition can wait until the teenage years.

The advantage of an open-air life

The most peculiar sight for a first-time visitor to Norway is often seeing baby strollers lined up outside of coffee shops and restaurants in the dead of winter. It's not that the proprietors want to keep the strollers out of their establishments but rather that the parents want to keep their babies—sometimes tiny babies—out in the cold, fresh air. (This can be a common sight in major cities in Denmark and Sweden as well, just a bit less striking because the weather is usually better.) If you ask a Norwegian mother why her baby is out in the cold while she warms up with coffee, she'll likely tell you that it's good for developing the immune system. If you ask a Norwegian father, he may tell you that it breeds toughness and comfort in the outdoors. Both are probably right.

Norwegians are proud of their love of the outdoors and the way they seem to withstand the elements better than people from places with better weather, which includes basically all the other 195 countries on earth. So much so that they have their own unique term for the lifestyle called *friluftsliv*. Literally translated, it means "open-air life," and it means just that. They're so fond of this term that it's one of the first things you see when you visit VisitNorway.com. Norwegians are born with skis (or hiking boots) on their feet, they spend as much of their youth and adolescence outside as possible, and once they have kids, it's time to get outside with those kids and do it all over again.

Walls and ceilings are reserved for school time, meal time, or sleep time, and even then, Norwegians like to do things differently. One concept that has garnered a bit of press in North American media recently is *udeskole*, or "outdoor school," which is often referred to as "forest schools." These are prominent throughout Scandinavia and are often boarding schools where parents send their kids for weeks at a time. Some are fully private, but like most things in Scandinavia, the majority receive a healthy amount of government support. The udeskole concept is even woven into traditional primary and secondary schools, typically with one day a week or every two weeks spent completely outdoors.

It's a type of free-range learning that is increasingly rare in the Western world, and it's part of what has created a generation of Scandinavians with an even greater love for nature, more independence, and an unbridled curiosity for the world they're ready to conquer. Kids learn by doing things, not by hearing things, and this often involves a lot of trial and error, which is something that has carried over to the Norwegian method of sport: Try something new, test it as much as possible, and keep only what works. As with the creation and evolution of the Norwegian method, a decades-long trial-and-error project can pay off when you are willing to try, err, and test a lot.

Unfortunately, it takes a bit more than happy, healthy, independent, and outdoorsy kids to build world champions. If that's all it took, Denmark (with the same population and even happier children) and

Sweden (with double the population and almost as happy of children) would be as or more dominant on the biggest stages of world sport. That's not to say that both countries aren't having a moment across a variety of sports, but as much as it would kill its two neighbors to admit, Norway is doing it a bit better.

Greatness that defies all odds

Norway's current golden generation of athletes comes from very different walks of Norwegian life and has very different upbringings. Erling Haaland and Jakob Ingebrigtsen, Norway's two most prominent athletes and the best in the world in the most competitive of sports, had rather cherished childhoods from a purely athletic perspective. Jakob's dad had him training as a professional at age eight, but Haaland had him beat by a few years. The son of a professional soccer player, Haaland was born in the UK while his father was playing in the English Premier League. He returned to Norway at age three and started in the development program for his local professional team, Bryne Fotballklubb, at just five. Haaland's upbringing, like the Ingebrigtsens, didn't exactly fall in line with the Children's Rights in Sports Act. As much as the Norwegian government and the Norwegian Olympic Committee aim to promote joy, independence, and a lack of competition among its youngest athletes, that doesn't stop plenty of parents from sending their children down a competitive and often joyless track from a very young age. It is a free country, after all.

For the most popular sports, and especially soccer, there are elite soccer academies that will treat kids as young as five as pros. This, of course, isn't unique to Norway. Nearly every major soccer club in Europe has development programs for kids from ages 5 to 16. Norway's programs just haven't received much attention because they've been terrible at men's soccer for as long as they've played. That all changed when Haaland burst onto the international soccer stage with Viking-like speed in 2020. All of a sudden, the international soccer media was paying attention to Norway like never before.

Endurance is no doubt a huge component of soccer, but it's not like Haaland grew up living by the Norwegian method like the Ingebrigtsens did. It's likely that the soccer star has never heard of double-threshold days and has probably only seen a lactate meter in a lab setting, if at all. But there are still remarkable similarities between the two athletes, who were born 59 days apart and were raised 11 miles away from each other. Both had demanding fathers, and Haaland's was a successful athlete himself. Both families also had the means to take what they needed from the academic and athletic structures provided by their home country and supplement where they saw fit—which was a lot. The travel restrictions placed on many young Norwegian athletes certainly didn't apply to its two biggest superstars, who got their first taste of international competition shortly after learning the fundamentals of their sport.

Haaland underscores the exceptional circumstance that two of the world's very best athletes—competing in sports that are arguably the most competitive on the planet—come from the same little dot on the globe and were raised there at the very same time. If you want your kid to become the very best in the world at something, the last thing you would want is for them to play soccer or run the 1500. There are at least a billion children right now who want to be either the best soccer player or the best runner on earth. Shooting, archery, equestrian, or any other sport with a high barrier to entry and a smaller competition pool offer exponentially better chances of reaching the top.

Triathlon might offer a narrow path to world-class greatness, and of course, Norway has served up another fascinating case study there. The fact that triathlon has a much smaller competition pool than running doesn't make Norway's climb to the top of the sport any less likely. It's still a very global sport, with 172 countries having national federations, Norway's being among the newest.

Kristian and Gustav grew up in opposite parts of the same place—Bergen—with the former sometimes catching jokes from the latter about growing up in the ghetto. (There is, of course, nothing in Bergen that could be deemed "ghetto" by any standard.) Kristian grew up close to the city center, in a more blue-collar part of town, while Gustav hails from the countryside, over one of the seven mountains that surround Bergen, which made for different lifestyles for the two future triathletes. But growing with slightly different means didn't stop either from finding the path to the top of their chosen sport in a country that makes that just a little bit easier. Or a lot easier, depending on where you're from.

It pays to live like a Norwegian

The real juice of the Norwegian method of training is in the Norwegian model of living. Both are focused on a holistic approach to the human body and mind, starting first and foremost with getting outside, being independent, and dealing with whatever comes their way. It works for a lot more than endurance. In addition to Haaland, Norwegian athletes have reached the top of the world rankings in some of the most competitive global competitions. Viktor Havland is consistently ranked in the top five in the world in golf, and Magnus Carlsen is the greatest to ever do the hardest game ever invented: chess. When it comes to both body and mind, the Norwegian model of life is generating extraordinary humans.

The fundamentals of the training programs of Kristian, Gustav, Jakob, or any of the other great Norwegian athletes aren't that much different from that of the best runners in East Africa, the best cyclists in Belgium, or the best speed skaters in the Netherlands. (The Dutch way outperforms even Norway on the speed-skating oval.) The differences between the Norwegian or Kenyan (or Dutch) methods are as subtle as the differences between polarized and pyramidal training. But the real ticket may be as much in living like a Norwegian as training like one.

The golden generation of Norwegian athletes (and chess grandmasters) owes a little bit to science but a lot more to the art of living well—something Norway has nearly perfected over the last century. The United Nations publishes something called the Human Development Index each year, which ranks all 196 countries on earth based on how healthy and wealthy its citizens are. Switzerland finished first in the 2022 rankings, which was most remarkable because it ended Norway's 14-year streak at the top. Now if this list were indicative of athletic performance, athletes from Liechtenstein would strike fear in the rest of the world. What it is indicative of is the foundation needed to mold healthy and happy children. The rest is a matter of good coaching and maybe a bit of luck when it comes to the right athlete finding the right sport—as was the case for Kristian Blummenfelt.

HUMAN

11

KRISTIAN BLUMMENFELT: THE TERMINATOR

Do not expect to make headway with a frail sailcloth.

—EYRBYGGJA SAGA

The most unique thing about Kristian Blummenfelt is that there's nothing particularly unique about him—physically, at least. His lungs and heart are a bit larger than the average human's, but certainly not off the charts among world-class athletes. At five feet, eight inches tall, his height is exactly at the worldwide average for adult males. His barrel chest and relatively short legs don't make for the most graceful of running forms, but there are no style points in his line of work.

If there is a single physical trait he holds above the rest, it's the two oversized pistons that powered him to the most impressive five-year stretch in the history of triathlon and perhaps all of endurance sports. His legs are heavily muscled for someone who spends so much time running long aerobic sessions, and they can flat-out fly at the end of a race, thanks in part to the relative ease they move through the first two disciplines. Kristian's legs are less that of a triathlon champion and more that of a Tour de France domestique.

That extra gear at the end of races—whether it's something unlocked in his head or his legs or, more likely, a mix of both—is also what makes Kristian the perfect athlete for the Norwegian method. His biggest threat to his training is himself, and he needs to be held back more so than most. Restraint is not something he does well in the areas of life he has deemed important.

Not one to stay in his lane

As with all the great ones, there's a bit of luck and happenstance involved in how a kid from Bergen—who really doesn't like Bergen—grew up rather quickly to become the best to ever swim, bike, and run. His upbringing was quite normal by Norwegian standards. Unlike Erling Haaland or Jakob Ingebrigtsen, he wasn't groomed by his parents to be an elite athlete from the time he could walk. Neither his parents nor his two older sisters were particularly athletic, and his dad was even a smoker throughout much of his childhood, something that is surprisingly still typical among Norwegian men. But like nearly every Norwegian, skiing and hiking became part of his life as soon as he could walk.

His first taste of competition came in the pool, and by age 11, swimming became his life, and winning races was all he could think about when he wasn't in the water. His only problem, his average physique, is a big limiter in the water. If you watched the medal ceremony for any of the swimming relays at the last Olympics, you probably saw 12 massive men well over six feet tall crammed together. Kristian's dream of becoming the best swimmer in the world probably wasn't going to come true, especially coming from a country that has produced exactly one world-class swimmer in its history.

Like most youth sports in Norway, the Norwegian model of swimming is a little more holistic than most, and Kristian's childhood coach frequently had him running outside as a means of crosstraining and getting a mental break from the pool. In the pool, Kristian's peers were just that. On dry land, he was leaving them in the dust without trying very hard. He got another taste of winning, and it turned out he kind of enjoyed running a bit. His coach signed him up for a local 10K at the age of 12, and when he finished in 36 minutes, it was clear that the kid who didn't look much like a runner had some serious talent for it and perhaps it was worth pursuing.

That's where the happenstance comes in. It's very possible that the very best football player ever born never played football. Or the kid who would have grown up to be the greatest hockey player of all time never laced up a pair of skates. It's even more likely that the would-be greatest triathlete on earth never did a triathlon, since the sport doesn't exactly have the lowest barriers to entry. It just so happened that, the summer after finding out that he was pretty good at both swimming and running, one of the neighboring villages announced it would be holding a triathlon, the very first in southwestern Norway.

Having just turned 14, Kristian was by far the youngest of the 32 men racing. It's not like he was competing against an elite field, and, like himself, most of them were competing in their first triathlon and had no idea how to do a transition. But he still finished so far ahead of the 31 grown men he was competing against that there is not another athlete to be seen in the photo of him breaking the tape.

If he comes, you must build it

Bergen is a small enough city that the news of a 14-year-old beating a bunch of adults in something as grueling as a triathlon made its way around town pretty fast. Fortunately for Kristian, Norway, and the course of endurance history, this news caught the attention of Roger Gjelsvik, the man responsible for pulling a lot of talent to Bergen's Tertnes Toppidrett. The only problem was, like every other toppidrett skole in Norway at the time, there was no program in place for triathlon. Norway didn't even have an Olympic triathlon federation. But a 14-year-old who was already better than adults at something he had yet to train for was too good to pass up. So much so that they began building a triathlon program—and eventually an entire federation—with just Kristian to start.

"There wasn't really anything physical we noticed about Kristian," Gjelsvik tells me as I'm sitting in his office, asking him to give me some detail of Kristian's physiology that makes him so much better than the rest. He clearly saw something at a very early age. "He's just different up here," he says, pointing to his temple. "Sure, the big lungs and capacity and all that help, but he was born with something you can't learn. It's in his blood. He's a Viking."

I asked Gjelsvik to clarify what he meant by the Viking comment. Was he talking about having a Viking mentality, or was he talking more about his bloodline and ancestry? Much to my surprise, he confirmed both, but more so the latter. He goes on to explain the history of Bergen—which I knew very little of at the time—and how it became a melting pot of cultures and races both during and after the Viking Age. According to him, it's why Bergensers look so much different from those in Oslo and other parts of Norway. Gjelsvik says, "We're just more mixed, I think. But not Kristian. As soon as I saw him, I said, 'That's pure Viking.'"

This could merely be a projection by Gjelsvik, who is quite proud of where he is from and the athletes he's had a hand in producing. All Kristian knows of his ancestry is that it's mostly Norwegian. Norwegians, for the most part, are less interested in their Viking past than many other Westerners are. He certainly doesn't think he has any ancestral advantage as an athlete, even if that's what Gjelsvik claims to have seen. What Gjelsvik and Kristian agree on is that what makes him different as an athlete is mostly in his head. His obsessive personality is perfect for both triathlon and the Norwegian method of training, since there are so many details for his brain to obsess over. Even when he's not training or learning about training, he's watching YouTube videos and listening to podcasts about triathlon. Some athletes go to great lengths to avoid reading what is posted about them on social media, but Kristian is the opposite. He absorbs the good and the bad (there's not a lot of bad) and extrapolates motivation from it all.

His obsessive nature has also built a world that takes the term *triathlon lifestyle* to new heights. He's not working triathlon into his life; he is creating a life that is only triathlon—or now, cycling. His life is essentially a training camp, spending most of his time in Font Romeu, France, or Sierra Nevada, Spain. During the three years between the last two Olympic Games, he estimates that he spent an average of three weeks per year in Bergen, which suits him just fine. He's quite open about the fact that he's not overly fond of the town he grew up in. He finds it a bit boring, the weather is more often miserable than not, and he spent the ages of 17–30 traveling to places he finds more inspiring. Simply put, the training and racing are better elsewhere, and that's all he wants to do right now. He's also quick to point out that the food is a little better in Spain and France.

With sponsors like Red Bull and Texaco, he's probably made more money at his age than any triathlete in the sport's history, but you'd never know it by visiting his apartment in Bergen—something he rarely does. It's as spartan as can be, with the lower level converted into a bike/run torture chamber that can be enclosed in plexiglass for heat training. The only decorations on the walls are medals and plaques that he's strung up haphazardly during his brief stays at home.

The big awards—the Olympic gold medal and the large Hawaiian bowls, or umekes, from his and Gustav's Ironman world titles—live at the pool, where he spends a lot of his time when he's at home. Those are always the rare easy weeks, when he gets to back off running and cycling, but there's little sense in backing off the swim. During his two-week "offseason" each winter, he can maintain a bit of that sharpness by swimming. There's not much consequence in the water for the life-long swimmer.

Consequence is something that Kristian thinks about often, and it's another element that has made this unique method of training fit his personality so well. It's a concept Olav preaches about constantly: If one of his athletes wants to go do a certain race or make a small change to the plan to accommodate life outside of triathlon, he reminds them of the consequence of each concession. Not that Kristian needs much of a reminder or has many compromises to make. He says that one day, he'll probably want a family and a home with some land—in France or Spain—but for now, he doesn't date or add the potential emotional stress of a partner to his budget. He has a few too many big goals before he's ready to start making accommodations for a partner or children.

It hurts more to lose

It is often remarked that great champions hate losing more than they love winning, and this is certainly true of Kristian, whose personal motto, which appears on custom T-shirts and other gear and stares back at him from the top tube of his racing bike, is "It Hurts More to Lose." He adopted the slogan after the 2016 Olympics in Rio de Janeiro. Just 22 years old at the time and ranked outside the world's top 25, Kristian was not expected to contend for a medal by anyone other than Kristian himself, whose hopes were likely inflated by a sixth-place finish at a World Triathlon Series event held in Stockholm a few weeks before the Opening Ceremony. But even he knew he couldn't run with British brothers Alistair and Johnny Brownlee, so his strategy was to stay with the lead swim pack at all costs, conserve energy in the lead bike pack, and then outrun everyone not named Brownlee to have a shot at the one medal that was really up for grabs.

Of course, everyone intends to stay with the lead pack in an elite triathlon, yet only a handful succeed. Rough waters at Copacabana Beach caused the 55-man field to split apart early, and Kristian found himself on the wrong side of it, exiting the water 15 seconds behind the leaders—close enough to believe he could breach the gap in the early stages of the eight-lap, 38.48 km bike leg. Instead he watched helplessly as the top men worked together to pull away from their desperate chasers, and by the time Kristian dismounted in the second transition, he was out of contention, though you wouldn't have known he knew that by the way he attacked the run, clawing his way past eight competitors over 10K to finish a respectable 13th.

Another athlete in his place might have celebrated—or at least taken the result in stride. But despite the fact that he had outperformed his ranking and despite having his whole career to look forward to, Kristian "went home devastated," as they say, tormenting himself for months afterward by replaying the race again and again in his mind, especially the critical moment when he lost contact with the lead swimmers, trying in vain to make it come out differently this time. His only solace was the realization that no future sacrifice he made in pursuit of winning could possibly hurt more than losing did. And not just losing—losing to 12 people, 10 of which weren't named Brownlee and were men he believed he should've beaten on the day.

"People love asking me how I'm able to dig so deep in races," he says. "My answer is always that the pain of not winning a race—or thinking that you could've gone deeper into the basement—that pain is what drives me in the race. I might have to suffer a lot for the next 10 or 15 minutes, but it's never as bad as the depression I'll feel for the next days, weeks, or months if I don't go there."

He certainly went there in his second Olympics, the pandemic-delayed 2021 Games in Tokyo, where Kristian drove himself into a state of otherworldly hurt that won him both a gold medal and legions of new fans who were moved by the visceral relatability of this barrel-chested, balding Norseman, blue Speedo showing through his sodden white racing suit, pulling away from a pair of graceful poker-faced whippets in the final lap of the run, his own face contorting like that of a bad actor pretending he's being eaten alive by tarantulas. Whatever other advantages Kristian might have enjoyed—altitude and heat training, lactate testing, and so forth—the average television viewer came away from watching his agonized surge to redemption with the conclusion that he won because of what he had going for him not below the neck but between the ears.

Rewriting the records and winning at every distance

If at his first Olympics Kristian learned that losing hurts more, at his second Olympics, he learned that the cure is winning, and he did a lot of that between his second and third Olympics. And when he lost, he lost mostly to Gustav Iden, which hurt less. Gustav's win wouldn't have been there without Kristian, just like a Blummenfelt world title wouldn't be possible without his best friend and training partner.

In the three years that represented the shortest Olympic cycle in history, Kristian won a World Triathlon Series title at the Olympic-distance triathlon; an Ironman World Championship eight months later when the race was held in St. George, Utah, for the first time; and an Ironman 70.3 world title on half of that course the following year. In three years, he won world titles at the three major distances, something that was so far beyond the realm of what was believed to be possible in triathlon since its inception.

With achieving impossible goals comes finding new, impossible tasks. It's the only way to keep feeding the monster. Between winning the Ironman title in St. George and finishing third in Kona in 2022, Kristian embarked on the "sub-7" project to see if he could finish a 140.6-mile triathlon in under seven hours if drafting and equipment regulations were removed. Using a team of professional cyclists on the bike and world-class marathoners on the run, sub-7 proved too easy for Kristian, who blazed around the Dekra Lausitzring racetrack in Germany in 6 hours and 46 minutes, more than 36 minutes better than the Ironman world record of 7:21 he set in his very first attempt at the distance in Cozumel in 2021.

The appeal of returning to the Olympic-distance triathlon to defend his gold medal in Paris was twofold: First, no one had ever done what he was attempting to do—go up and down in the distance and essentially hold all of the sport's biggest titles at once. And two, not defending a gold medal was out of the question. He reveres the sanctity of the Olympics above all else and knows it's better to go down swinging than not put up a fight.

Paris sets the stage for a pivot

The fight was valiant, even if the result was one of the most disappointing of his career. He knew that returning to the top of the Olympic podium after three years of long-course dominance would be harder than anything he'd ever attempted, but he never thought he'd finish so far from it.

After surviving a chaotic swim in Seine River, Kristian emerged as the protagonist on the bike, bridging the first chase pack up to the leaders with relative ease and then driving the pace at the front of the group, occasionally ordering an athlete behind him to take a turn a turn at the front. He'd come a long way from being one of the youngest and most inexperienced athletes at the 2016 Games.

But the result was similar. As in Rio, there were two men in Paris on a different level on the run—Great Britain's Alex Yee and New Zealand's Hayden Wilde—and coming out of transition both were gone in the blink of an eye to settle gold and silver. (Yee won gold with a dramatic last-second comeback.) Yet once again Kristian wanted—and expected—to be in the mix for the one medal that was truly up for grabs. He ran with that intention, maintaining contact with the small group of men with a shot at bronze, but he couldn't match the accelerations on the final lap, and was the 12th man across the finish on the Pont Alexandre III bridge.

There was more disappointment and less depression than Kristian's 13th-place showing in his Olympic debut. The allure of a nearly impossible goal carried an undeniable risk of failure. Over his career the only cure for such an outcome has been to move onto the next impossible task as swiftly as possible.

The response to Olav's announcement just before the Paris Olympics that he and Kristian would likely shift their focus to professional cycling in 2025—with hopes of "taking jerseys" by 2028—was as they expected and hoped. It was ridiculed as impossible, with critics saying Kristian was "throwing away the greatest triathlon career ever" and entertaining "a long pipe dream that won't end well."

It's all fuel for the fire that isn't likely to stop burning any time soon, even as Kristian and Olav reassessed their goals in the aftermath of Paris and announced that they would likely pursue triathlon gold (again) in 2028. A move to cycling could very well mean giving up a chance to go down as the greatest ever to do triathlon. It's a consequence to be considered. Whatever Kristian decides to do in the final chapters of his career, it's likely to be something everyone but he and his coach considers impossible.

HUMAN

12

GUSTAV IDEN: THE JESTER

One's back is vulnerable, unless one has a brother.

—GRETTIR'S SAGA

Aside from spending a great deal of time considering the consequences of every action, Norway's other triathlon world champion has very little in common with his training partner and best friend, at least in terms of personality. They find themselves on opposite ends of the introvert-extrovert spectrum, and perhaps that's what's helped create one of the most successful training microcultures in recent history. You can't just throw any personalities together and hope they'll push each other just right. Chemistry can't be forced, and this one works like no other training kinship in endurance sports before.

It's a unique relationship. Olav Aleksander Bu is somewhat of a father figure to two friends who have become closer than most brothers. And the brother who is just three months younger is undeniably the jester of the trio. While Kristian and Olav are perfectly content sitting in silent thought for minutes on end, Gustav makes sure every break between intervals or dull moments during travel is filled with jokes, laughter, and general observations about the world. He doesn't do silence well and would probably be a good stand-up comedian.

During a swim session in Bergen—a rare time that the two world champs are training with a larger group of up-and-coming Norwegian triathletes—the seconds between 100-meter repeats are filled with playful cut-downs for Gustav's friends, Kristian included. He's delicate and funny enough that everyone is usually laughing too hard to be offended. During one break, he tells Casper Stornes—who in his own right is one of the best triathletes on earth—that he will never become a world champion because he has a serious girlfriend. A few repeats later, he has another thought and tells Stornes that he plans to buy the house he and his girlfriend were considering purchasing because he is a world champion and has more money. He adds that he'll pay more than it's worth just to keep Stornes from buying a house with his girlfriend, which would further hinder his chances of becoming a world champ.

After swim practice, I spent 30 minutes interviewing Gustav about Kristian, whom I'm in town to do a story about for one of his sponsors. This amuses Gustav to no end because a 30-minute interview about his friend is 30 minutes that he gets to make jokes about him. He continues on the girlfriend bit, noting that it's harder for him not to have a girlfriend, since Kristian is so awkward around women that it just takes care of itself.

He asks if I'm interviewing any of the other athletes about Kristian, and I say I'd like to talk to Stornes because Kristian had mentored him a bit. He then proceeds to teach me a word in Norwegian that he claims is Stornes's nickname and says I should call him that. I know it's a trick, but I play along when I see Stornes a few hours later. I repeat the word after much practice on my way to see him, and he just shakes his head and says, "Fucking Gustav." A few seconds later, I learn that I'd just called one of the best triathletes on earth "Pussy Face."

The perfect setup for two out of three sports

Few athletes at Gustav's level can hold on to a degree of frivolity as he seems to all the time, and it's what has made him the perfect yang to Kristian's often intense ying. The two humans aren't just different in the head; their physiology is quite disparate, and that's part of what has made Olav's success in training both of them similarly but individually even more remarkable.

Beyond a shared slate of coaches, one of the only similarities in their respective childhoods was that Gustav also had two older siblings—one brother and one sister—and, as is often the case, the youngest one becomes particularly competitive to keep up with the older, faster, and stronger siblings. On the spectrum of parents pushing their kids toward elite athletics, Gustav fell somewhere in the middle. He wasn't like Kristian, whose parents really didn't push sports on him at a young age, nor was he like Jakob Ingebrigtsen, who was primed for elite sport by age eight. His dad was an avid cyclist who coached one of the biggest cycling clubs in town. Both his older siblings were already riding with the club by the time he learned to ride a bike, and living in the countryside outside of the city, it wasn't hard for Gustav to fall in love with life on two wheels.

Gustav also had plenty of talent as a runner and routinely dusted every kid in his school during recess or when they'd play sports like soccer. He knew he had above-average speed, but he grew up surrounded by cyclists telling him that running was a surefire way to ruin one's cycling legs, and almost no one in Norway had even heard of triathlon in the early 2000s. As is often the case when kids become teenagers, he lost interest in his childhood hobby at age 13, mostly because riding by himself or with his siblings was getting boring. He found he'd much rather be running with his peers. Or more specifically, running away from them. By 16, he won a gold medal at the world junior cross-country championships and was invited to try out for Norway's brand-new national triathlon team, which at the time had exactly one member: Kristian Blummenfelt.

Unfortunately, Gustav would have to wait an extra year to join his soon-to-be best friend at Tertnes Toppidrett and the new Norwegian tri squad because he was told his swimming was so bad that it was unlikely he'd ever make it on the international level. That motivation and 12 months were all he needed to learn how to swim somewhat competently and get accepted for his second year. His swim progression came nearly as quickly as his run had a few years earlier, aided by the fact that he had Dr. Madsen and his very nice pool just a short ride down the mountain. From the moment he joined Kristian, the two hit it off as training partners across all three disciplines. Kristian had a lot to teach Gustav in the water, and Gustav had nearly a decade more experience racing a bike. The run was a relatively even playing ground, with both demonstrating world-class speed as teenagers.

The 10-year head start Kristian had in the water made it impossible for Gustav to keep up with the competition for a few years. (Top athletes often compete in a junior category in World Triathlon Series events from ages 15 to 19 before progressing to either under-23 or elite racing.) Kristian spent just one year competing as a junior—at age 16—before it was clear that he was ready for the elite level. His progression was lightning-quick, thanks in large part to his swimming background. Gustav had to be considerably more patient, spending two additional years racing as a junior while also putting in six days a week in the pool to get his swim to the point that he could compete at the highest level.

Gustav made his national team debut at a Junior World Cup in the Netherlands in 2012, finishing 33rd out of 70 athletes. It was a humbling experience for a kid who'd become accustomed to winning cycling and running races, and about the only saving grace was that he'd beaten his older brother, Mikal, by nearly a minute. But even his slower and bigger brother had managed a respectable swim split. Of the men finishing in the top 40 at that race, Gustav was the only one who needed more than 10 minutes to make it through 750 meters. His swim split was 10:26. Kristian, who finished third and exited the water in the lead pack, finished the swim in 9:13.

After two more years of work both at the pool in Bergen, swimming alongside Kristian, and with his home club in Åstveit, he swam 39 seconds faster at the very same race, which allowed him to use his considerable power on the bike to bridge up to the lead pack, eventually going on to finish third. One more year of work and his swim was unlocked, exiting the water in Holten in 2015 in 9:09—even faster than Kristian had swum in 2012—en route to winning his first international race. The hand-wringing over Gustav's swim stopped, and it was time to join his training partner in the elite ranks.

Home is where *hygge* is found

Like most Bergensers, Gustav is very proud of his hometown and returns from his world conquests as often as he can. However steep his ramp to the pro ranks was, it hasn't affected his playful disposition.

Maybe it is a by-product of growing up over one of the seven mountains in the region—in the countryside, where the days are a bit less bleak and the riding is a bit better. Bergen is objectively beautiful, and its ancient harbor is a United Nations Educational, Scientific and Cultural Organization (UNESCO) World Heritage Site. But there are plenty of days—weeks, even—when you could be sitting in the middle of the harbor and not even be able to see the vibrant buildings UNESCO deemed so important due to clouds and fog. It's not the sunniest place on earth, and that's certainly not for everyone.

It's definitely for Gustav, who spends just slightly more time in Bergen because he still relishes and respects the importance of being away at training camp. But even when he's not in Bergen, he embodies the spirit of his town in a way Kristian certainly does not. There's a Scandinavian sentiment known as *hygge*, which is a Danish word that is used by Swedes and Norwegians alike. The closest way to describe it in English is contentment with one's surroundings and everyday life. It also captures Norway's tendency to shy away from boastfulness. They're more likely to say that they're content—not necessarily happy—and it's a peace to be enjoyed within, not broadcast outward.

Hygge isn't so prevalent in Bergen, which some consider to be the black sheep of Scandinavia. Oslo, Copenhagen, and Stockholm—the other three major cities—are all close in proximity and share plenty of cultural correlations. Bergen is very much its own place, and so are many of the people who call it home.

"We like to say we're from Bergen first and Norway second," Gustav explains. He's not kidding. At the stadium for Sportsklubben (SK) Brann, Bergen's soccer club, the Norwegian flag and city flag fly at the same apex. And they're boastful, even when it's a bit of a joke: "Our soccer club is bad. So bad that we're often in the second division in Norway. That's really sad for the second-biggest city in the country. But we still say SK Brann could beat any soccer club in the world if they had the chance. We don't really believe it, but we kind of do. We have a weird belief in ourselves that way."

The role of belief and a little luck

It's that inner belief that led to the moment that shocked everyone in the triathlon world except Gustav at the 2019 Ironman 70.3 World Championship in Nice, France. Competing in just his third Ironman 70.3 race at the age of 23, Gustav went from a relatively unknown athlete to a world champion over the course of 3 hours and 52 minutes on the Cote d'Azur. The man finishing in second place, three minutes behind him, was none other than the then two-time reigning Olympic champion Alistair Brownlee of Great Britain. Kristian was fourth, nearly seven minutes behind. Gustav had beaten the hell out of the guy who had just won the last two Olympic gold medals in dominant fashion and the guy who was about to win one in an equally dominant manner. Over a harder course and longer distance, the Olympic champions were no match for Gustav.

News of Gustav's world title becoming big news in Taiwan is yet further evidence of his quirkiness. While doing a training camp in Japan a few months before the race in France, Gustav noticed a blue hat with yellow embroidery lying on the side of the road during a run. He picked it up and saw it had something written on it—he only knew enough to know it wasn't Japanese—and put it on and kept running. He liked the way it looked, it was comfortable enough for running, and it would do fine on race day. Within a couple days of winning in France, he began to get calls from the media in Taiwan asking for interviews about the special hat. It was only then that he learned it was from a famous Buddhist temple near Taipei. He told reporters in Taiwan that it was his "lucky hat" and has become something of a local celebrity ever since. He has also worn some version of it in every race since 2019, although the original mysteriously went missing after the 2021 70.3 World Championship in Utah. Someone at Ironman—which was owned by a Chinese billionaire at the time—asked him to remove it for an interview, which he did, and that was the last it was seen. A replacement sits with his world championship trophies at the pool in Bergen, and meanwhile, the temple in Taiwan happily sends him replacements whenever he needs them. They've even made more run-friendly versions just for him. It's one of the strangest sponsorships in sports, with Gustav getting nothing out of it except free hats and, perhaps, a bit of luck.

A cruel twist of fate

It hasn't been all laughs and luck for the man who seemingly waltzed his way to his first world championship at 23 and had everything going right. Even the onset of COVID-19 in 2020 couldn't deter Gustav's rocket trajectory when he won the only major triathlon held that year at Challenge Daytona. He followed that up by finishing seventh at the Tokyo Olympics. As much as he wanted to share a podium and medal with his friend, a top-10 finish was remarkable for someone who was clearly better suited for the longer stuff. Just six weeks after Tokyo, he won his second world title in St. George, as Kristian struggled on the bike and finished near the back of the field. At his full Ironman debut two months later in Florida, Gustav won on the heels of a 2:34:50 marathon, which was the fastest Ironman marathon ever at the time.

Things were perfect going into 2022, when he would have two chances at winning his first Ironman World Championship: The 2021 Kona race had been moved to St. George in May of 2022, with the regular race in Kona taking place that September. St. George, with its endless hills on the bike and run, seemed to be the best bet for Gustav, who had just won his second world title there. It was shaping up to be the best year of his life—that is, until his parents called while he was at training camp with devastating news: His mom had metastatic breast cancer, and it was advanced. The year ahead would involve a lot of treatment and surgeries for her, but she didn't want it to deter her son's plans. She was insistent about that. She wanted him to chase his dreams even harder than before.

And he did. His mother's battle and declining health throughout 2022 were both a distraction and a motivation. There was energy to be harnessed in all the emotions, and training went almost perfectly throughout the year. Unfortunately, an illness forced him to sit out the first of the two Ironman World Championships, instead watching in St. George as his friend won yet another major title. There was disappointment in knowing that he was as much on form as Kristian—perhaps even better. It was Gustav's race to win, but he suggested that perhaps it was a victory Kristian needed even more.

"I think it's all about winning for Kristian in a way that it isn't for me," he says. "It's like it's all he thinks about—that moment when he crosses the finish first. I'm maybe a bit different. I can be happy or OK with a race I don't win in a way I don't think he can."

He can also enjoy a win as much as his marginally more successful friend, especially when it means something bigger. At 26 he came to the realization that when he responded to the cannon's boom in

Kona, his mother would likely be watching him compete on one of the sport's biggest stages for the last time. One of the many top contenders interviewed by veteran triathlon journalist Bob Babbitt on the eve of the race, only Gustav seemed as relaxed as he was determined and as eager as he was confident.

"So what's the goal tomorrow?" Babbitt asked off-handedly at the close of a 20-minute Q&A.

"Isn't it obvious?" Gustav asked, shrugging.

If it wasn't obvious then, it was the next morning when Gustav executed the closest thing to a perfect championship Ironman race that the sport has ever seen. Intent on outsmarting the competition, he performed no heroics in the swim, conserving energy near the back of a 19-man lead group. The race remained bunched up at the front throughout the bike leg, which Gustav must have seen as an advantage, having also told Babbitt that he's at his best in highly tactical races, where the victory goes to whoever makes the best decisions. And as a result of his good decisions, Gustav burned just enough energy to start the marathon alongside his biggest threat—Kristian—and six minutes behind France's Sam Laidlow, who may have burned too many matches in eviscerating the bike course record.

Trailed by a doomed Max Neumann of Australia, who held on for fourth place, the Norwegians ran side by side and cut into Laidlow's lead with mathematical exactitude, clicking off 3:40 kilometer after 3:40 kilometer. But approaching 19 miles, Kristian faltered, no longer able to do his share of the work. Taking his cue, Gustav glided away from his overmatched training partner and closed the remaining distance to Laidlow within three more miles, while behind him, Kristian dug deep to fend off a stubborn Neumann. When the inevitable happened, Gustav greeted Laidlow with a collegial shoulder pat, and the Frenchman offered the Norseman his hand, which he took before surging ahead to the biggest win of his life and a new record on the sport's most legendary course of 7:40:24—a time thought impossible by previous generations of Ironman champions.

Coming up short of the Olympic short course

It bears mentioning that the man whose record Gustav broke, Jan Frodeno of Germany, is considered to be the greatest male triathlete of all time and the first ever to crown himself both Olympic and Ironman champion. That legacy is one reason Gustav made the seemingly preposterous decision to return to short-course racing and chase the Paris Olympics after so much success at the longer distances. Unlike Kristian, there's no Olympic medal to defend, and that's part of the problem. Even though Gustav took down Frodeno's Kona record—one of the shiniest records in all of triathlon—he hails from a country obsessed with Olympic medals, and it's been his dream since he started riding bikes as a young boy. He was well aware that his chances were slim and that he was likely giving up multiple long-distance world championships over the course of the two years he rededicated himself to the Olympic-distance triathlon. That could be a couple million dollars in lost incentives and sponsorships.

But Gustav never really looked at it that way. If he shares just one more thing in common with his friend, it's that money is not the motivation. It's a result, not the end goal. For Kristian, the money is a result of his endless need to win and keep winning. For Gustav, racing is a commitment to the process that he has enjoyed tremendously since he was a kid. His goal is to keep getting better each week and continually learn more about himself and his sport. It's that process-driven mindset that has helped him through the first and only rough patch of his career in the aftermath of his record-setting win in Kona. A nagging Achilles injury throughout 2023 and the passing of his mother led him to step away from racing and back off run training. But in his return to competition and the lifestyle he enjoys so much, Olav says he sees Gustav as a more mature and stronger athlete than ever before. He's confident that the second half of his career will involve even more success, even as he watched Kristian and rising star Vetle Bergsvik Thorn compete for an Olympic medal in his stead.

He's matured quickly as an athlete and has learned a lot about himself as a person over a couple of trying seasons, but both he and his coach have only belief that he'll get back to the top of the sport. There's no room for doubt when you know exactly what to do. His parents taught him to love the process—and the outdoors—at a very young age, and he knows there's nothing unique about him or his situation that makes him the best in the world. It's just about doing and enjoying the process better than anyone else.

"People always ask what's so special about Bergen that's made so many good triathletes," he says. "It's nothing special. It's a bit of luck and good timing with me and Kristian. The right people got together at the right time, and we set these very big goals from a long way out."

SCIENCE

13

FEEL THE HEAT: THE EFFECT OF CORE BODY TEMPERATURE

Et tre med sterke røtter ler av.

A tree with strong roots laughs at storms.

—NORWEGIAN PROVERB

In the days following Kristian's Olympic victory in Tokyo, a number of media reports from around the globe were followed with stories wondering how someone from one of the coldest places on earth had handled the heat so much better than everyone else. Some talked about his unique white race suit, which was essentially see-through except for the blue briefs he wore underneath. Others focused on the heat-training chamber he had built in the lower level of his home in Bergen—though there was little mention of how little time he spends in Norway. When asked by the *Japanese Times* how he had handled the Tokyo temps and mugginess without wavering, he just shrugged and said he was "disappointed it wasn't hotter."

It wasn't just Kristian. All three Norwegian athletes finished in the top 11 and seemed rather unfazed by the conditions. It took a lot of time away from Norway to pull off such a feat, including trips to places significantly hotter than Japan. Norwegian athletes have been seeking higher temperatures for almost as long as they've been seeking higher altitudes because it's nearly impossible to get either outside of a lab setting at home.

Heat acclimatization certainly isn't uniquely Norwegian, and Norway's coaches and athletes were hardly the first to experiment with it. But their triathletes, in particular, have used some distinctive methods to prepare for blistering conditions, which is fitting given that in the sport of triathlon, Hawaii's lava fields are home to the Super Bowl of the sport and effectively serve as a crucible that makes or breaks champions.

The early days of heat training

Thanks to triathlon's longtime marquee event—Kona—taking place on an oppressively hot island smack-dab in the middle of the South Pacific, triathletes have been experimenting with some eccentric training protocols since the first Hawaii Ironman in 1978. For a long time, the elite Ironman's approach to heat acclimatization prior to Kona was fairly simple and fairly effective: Arrive on the Big Island as many days as possible before race day and get out and train at the times you'll be racing. Key to this strategy: Don't you dare turn on the air-conditioning unit wherever you're staying. Admittedly, it was a miserable experience for both the athlete and anyone traveling with them, but that was kind of the point. Without having any idea what physiological adaptations were happening (i.e., increased sweat rate and blood volume), these early triathletes were just trying to get as close as possible to acclimating to the conditions before the race cannon went off.

One of the first and most famous instances of specific heat training for the Hawaii Ironman was employed by six-time champion Mark Allen, who frequented a sweat lodge during his spiritual retreats with a Huichol shaman. He was seeking something in addition to heat acclimatization during his stays in a sweat lodge, which likely put him on the verge of heatstroke. Whether or not he found it, it appeared to be an effective strategy for adapting to the conditions in Kona. It would be difficult to find a national sports body willing to fund a scientific study investigating the merits of sweat lodges for heat adaptation, so all we have is an n = 1 study of Allen. Prior to the retreats and sweat lodges, he found himself fading late in the run as the legendary Dave Scott—a six-time Kona winner himself—reigned over the Big Island. Once he began studying (and sweating) with a little shamanic guidance, it was like he was a different athlete, and he put the Kona record books out of reach for an entire generation. Now, there was way more to Allen's victories than sweat lodges and extreme heat training, but his success got a lot of other triathletes wondering if they should be doing something more prior to arriving on the Big Island to get ready for the unforgiving conditions.

The effects of thermal load

Relative to other important aspects of endurance performance, heat training is still a fairly fresh science that is just beginning to be understood and effectively applied to sports. Given the degree to which heat affects energy metabolism, it might have been better as one of the first components of exercise physiology that researchers tackled. Here's a rough overview of how it works: For each liter of oxygen a person takes in, five calories (or kcals) of energy are produced. Of those five, only

The Value of Heat Training

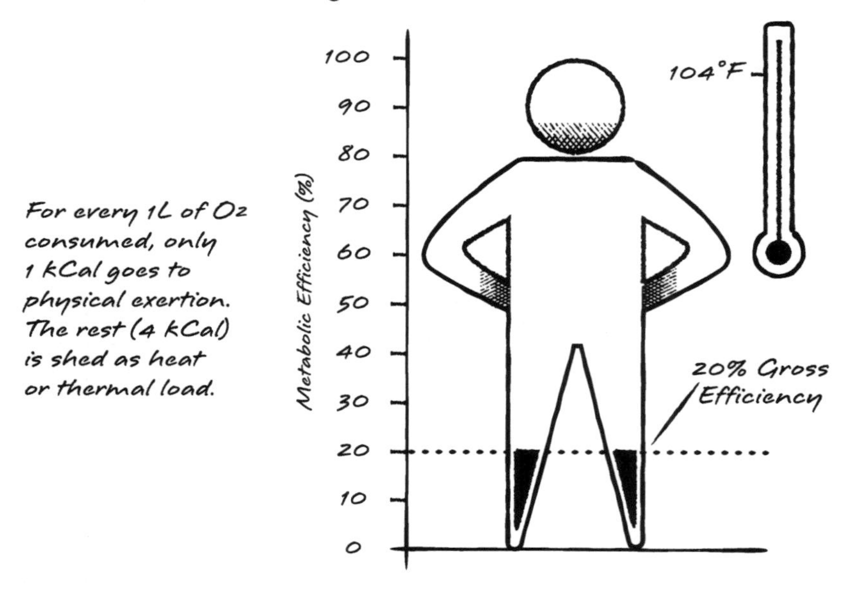

The metabolic system presents a significant limiter—your body is roughly 20% efficient. If you can build a bigger furnace with capacity for an internal body temperature above 104° F, you can extend this efficiency beyond the typical human limits.

one goes toward physical exertion, which, in the case of an endurance athlete, results in forward motion. The other four calories are converted to heat, which the body has various means of handling. For example, if a cyclist is producing 1,000 watts of metabolic energy, 200 of those watts are directed at the pedals and cranks to produce the number on the power meter, and the other 800 watts are effectively lost to what is frequently referred to as *thermal load*. The bottom line: The gross efficiency of the human body is 20 percent, and there isn't a whole lot that can be done to change that fact. It's why coaches and nutritionists have spent decades trying to ensure that calories consumed are converted to energy as close to that 20 percent rate as possible, and it also explains why maintaining a steady core body temperature has emerged as their primary objective.

A 2019 study of 15 elite cyclists from Spain's Basque Country found that gross efficiency dropped one percentage point for every degree (centigrade) gained in core body temperature. That's a lot when considering the body is, at best, 20 percent efficient to begin with. An increase of a couple of degrees can lead to a drop of 20 watts or more, which is often the difference between breaking away to victory or fading back into the peloton. For the longest time, sports scientists, coaches, and athletes approached this problem with a rather fixed mindset: How do we keep core body temperature down in hot environments so that we don't lose any of this precious gross efficiency? Much of the research now, especially among Norway's most prominent triathletes, is focused on maintaining that 20 percent gross efficiency even as core temperature increases into ranges that were previously believed impossible—or even dangerous.

One of the biggest obstacles to researching how an athlete's body performs in heat is that race conditions can't be readily simulated in a lab, and further complicating things, there are laws governing how hot lab settings are able to be (usually around 104°F). Longer events, like a marathon, cycling road race, or triathlon, are basically impossible to simulate in a lab setting. The only way to know what's happening to the body during those events is to test in real time, which has been a limited course of action until recently. Measuring an athlete's temperature at the skin level doesn't tell you anything about core body temperature, which is the limiting factor that ultimately causes an athlete to slow down or go unconscious.

The effect of duration on core body temperature

As with many elements of endurance technology, like power or aerodynamics, cycling has been at the forefront of studying heat. One of the most groundbreaking studies came out of the 2016 Union Cycliste Internationale (UCI) Road Cycling World Championships in Doha, Qatar, where temperatures averaged 100°F during the races. Cycling has some of the longest and hottest days of racing of any sport, so it's understandable that cyclists have been eager to find ways to "hack" the heat. As it turns out, one of the most notable beliefs in endurance studies ended up being disproved by the Doha study. It was long thought that the longer the event, the higher the core body temperature could go. It makes sense, but in Doha, researchers found the opposite to be true.

The study, published in the *British Journal of Sports Medicine* in 2018, examined how 40 athletes at those world championships swallowed ingestible heat pills to record the temperature in their guts during the road races and time trials. Truth be told, the pills had to then be retrieved in order for the data to be collected, so it wasn't real-time information, but it did provide data for the entire race. Those little pills debunked two previously held tenants of endurance:

1. The temperature ceiling for athletes before heatstroke or slowdown is not 104°F, as once thought.
2. In fact, the longest races do not cause the greatest increase in core body temperature.

Case in point, the highest number recorded was 106.7°F in a female time trialist who was competing on a day with air temperatures of

99°F (the hottest day of competition). It was an extraordinarily high number from the shortest race, shattering the 104°F ceiling that was previously thought to limit well-trained athletes and demonstrating that short and fast race efforts can produce more heat than slower and much longer ones.

But the pills had their limits. Measuring temperature in the gut is much more precise than anywhere on the surface, but most of the athletes were consuming cold liquids before and during competition. Again, independent variables are hard to come by in the real world. It's not like the researchers could've asked the world's best cyclists to hold off on hydrating in Qatar. It's safe to say that hydrating affected the internal temperature readings, and determining to what degree it did was beyond the scope of possibility. A more precise reading would take a more precise—and invasive—method. And a few athletes committed enough to winning that they might just do it.

Upping the heat to raise the ante

Enter the Norwegian triathletes. In the buildup to the Tokyo Olympics, heat became a primary focus for all 110 triathletes competing, but the three Norwegian men—and Kristian in particular—handled it better than the rest. It turned out that the most disruptive Olympic Games in history provided the perfect opportunity to dial in their approach. Both in February of 2019 (prior to the Games being pushed back a year as a result of COVID-19) and in 2020, the Norwegian team traveled to Thailand with temperature-sensing pills in hand, but these were designed to be inserted into the rectum. The thinking was simple: Measure the athlete's true core body temperature with as little skew as possible from the nutrition being consumed.

During a two-week stay in Thailand, all three of the Norwegian Olympians performed an Olympic-distance triathlon at race pace on consecutive weeks with the pill inserted and did the same at the Olympic Test Event in Tokyo in 2019. Between two camps in Thailand and the race in Tokyo, they now had five "races" of heat data to mine, and the findings were as intriguing as those from Doha—and likely even more accurate. All three athletes demonstrated an ability to maintain velocity at over 104°F of core body temperature, with Casper Stornes setting the high mark en route to finishing second at the test event with a temperature of 105.8°F. That reading came at the end of a 30:27 10K in 92°F heat while trying to sprint for the win. It certainly isn't a sustainable number for any endurance athlete, but it had them more convinced than ever that in order to reach their highest gear in hot conditions, being able to get the body warmer might be even more important than keeping the core body temperature down.

One of the biggest takeaways from the Norwegian sample of five races further underscored what the 2016 Doha study first revealed: The athletes who are able to get the hottest are often the ones who perform the best. At the women's time trial in Doha, the three athletes with the highest recorded temperatures were the three on the podium. At the Olympic Test Event, Stornes surged away from both his countrymen with a core body temperature they couldn't match. Gustav finished fourth at the same event—one of the best results of his short-course career—and measured only half a degree less than Stornes. (Kristian crashed at the end of the bike leg and did not start the run.)

Since collaborating closely in heat preparations for the Tokyo Olympics, Arild and Olav have set out separately with their own group of athletes, but heat remains a steadfast focus for both coaches and likely will for the foreseeable future. The races aren't getting any colder, and you have to go seek heat when you're from Norway.

"Where we are now is building a training program to push the limits of how hot athletes can get, and it's a bit controversial because there's a lot of risk in it," Arild explains. "But it goes back to control and doing things slowly."

There was one more big takeaway from those five datasets the Norwegian team had gone to such lengths to produce, and it's a good strategy for triathletes of all abilities: Getting too hot on the bike makes it impossible to get hot enough on the run for peak performance. It was a valuable lesson, particularly for Kristian and Gustav, whose strength on the bike allows them to be the instigators and pace drivers. Two years later, when it mattered, Kristian understood the potential cost of being the aggressor and instead let some of his biggest rivals get too hot on two wheels. (They were not racing with rectal pills at the Olympics. At some point, comfort has to take priority over data.)

From the moment Olav began working with world-class triathletes, he's been obsessed with monitoring core body temperature. It's one of the areas he readily identified as "low-hanging fruit," meaning the competition hadn't spent a whole lot of time maximizing it, so he felt there was a big competitive advantage that could be mined. It's also one of the aspects of Kristian and Gustav's training that they've been more open to sharing because, for a long time, the methods used were so invasive that it would prove difficult for most competitors to seamlessly incorporate them into their training plans.

Body temperature sensors

Prior to 2020, the best methods of core body temperature monitoring for an athlete in competition or during training involved swallowing pills like the cyclists in Doha or inserting suppositories like the triathletes in Thailand or Tokyo. In 2019, a Swiss company named CORE developed the world's first noninvasive, real-time core body temperature monitor, with Kristian and Gustav serving as beta-testing guinea pigs, along with the Quick-Step pro cycling team. Both were so impressed by its efficacy and the fact that they no longer had to swallow or stick anything anywhere that they were two of the company's earliest investors, alongside Olav. Their involvement on the ground floor signaled a meaningful endorsement that reversed the typical sponsorship arrangement. That's how much they believe real-time core temp monitoring can elevate endurance training.

The CORE sensor aims to do what even suppositories and pills can't by measuring the true temperature near the internal organs in the torso without any interference from the other liquids and solids sloshing around an athlete's insides. But more importantly, it's a tool that can be used every day instead of on select occasions, allowing athletes and coaches to more closely monitor how much heat is being applied and how the body is adapting. CORE's patented sensor claims to capture the amount of heat flowing into and out of the skin and then uses a continuously updated AI algorithm to accurately compute the temperature surrounding the internal organs. That number is then transferred *in real time* to any paired ANT+ or Bluetooth device.

Real-time heat sensors have the potential to be the biggest technological breakthrough in endurance training since power meters, with perhaps even more benefits to offer. Cyclists, in particular, have long joked that heat is the poor man's altitude or EPO. They knew from experience that a lot of heat training produced similar physiological changes as an altitude camp—or even a cycle of EPO. It's more than just increased sweat rate and generally feeling cooler in hotter conditions. Increased blood plasma allows the body to maintain hydration levels without compromising other bodily functions—like digestion—which is crucial in ultraendurance events. The heart can pump with higher stroke volume and produce more cardiac output after sustained heat training, with CORE claiming as much as an 8 percent boost in stroke volume with the right heat-training protocol.

The concept that CORE (and Olav, one of its earliest investors) aims to teach the worldwide endurance population is what they call "thermal load," and it's different from the term that is usually applied to energy metabolism (e.g., the 800 out of 1,000 watts that are "lost" to heat, discussed earlier). In this instance, thermal load refers to the amount of time spent in the ideal heat-training zone, which is determined based on the athlete's goals and experience. Ideally, like any zone or curve, it moves with improvement. To keep things safe and avoid lawsuits, CORE's protocol recommends only two or three heat-focused sessions per week, with 45–60 minutes spent in the athlete's specific heat zone. The CORE app then tracks this load throughout the week, month, year, or even longer.

That's the piece that CORE, Olav, and the athletes who used this emerging technology in the buildup to the 2024 Olympics have been willing to share so far. They experimented with some new methodologies for Paris, but unfortunately for Kristian and Olav, heat wasn't much of a factor on race day. Olav believes there's still plenty of low-hanging fruit to be harvested when it comes to heat, and he suggests that the science of heat training is 20 or 30 years behind altitude.

But, along with other coaches on the cutting edge of it as well as the engineers at companies like CORE, he's confident that they can close that gap expeditiously. Norway—and the rest of the world—has a lot to learn about how the body handles getting hotter. Like altitude training, which offers significant benefits down the mountain, the new science of heat training will affect the way the best athletes prepare for races in any climate.

HUMAN

14

JAKOB INGEBRIGTSEN: THE CHOSEN ONE

To take up great resolutions, and then to lay them aside, only ends in dishonor.

—KING OLAF TRYGVISSON'S SAGA

There is no better embodiment of the Norwegian method than the Ingebrigtsen family and no better example of its merits than the youngest of its runners, Jakob, who has been doing some version of the method since he was eight. He was primed for the life of a world-class runner, but he had no chance of a normal childhood.

Not that Jakob or any member of his family has ever been interested in normalcy. With Henrik 10 years older and Filip 8 years ahead, Jakob's early days were spent chasing much older brothers who were already on the fast track to the Olympics. Gjert, their father, had ambitious goals for his three children—at least the three that he sent down the running path. Something that's not often mentioned is that there are seven children in the family—three more boys and one girl.

Whether it was at their mother's or father's behest, the other four were spared from a rigged childhood of running as well as the fame and fortune that has come along with it.

A track star is born

Jakob's rise to becoming the most famous person in Norway not named Erling Haaland began at age 10, when he was one of a dozen athletes featured on a Norwegian TV program profiling the best young athletes from around the country. Jakob was so charismatic and there was so much interest in his very fast older brothers that the show's producers decided to make an entire reality show about them two years later, and *Team Ingebrigtsen* was born, becoming one of the more popular TV shows in Norway and eventually gaining worldwide appeal on YouTube. He certainly wasn't the first person to grow up with all the demands that come with being the offspring of a demanding parent with lofty goals. But there seems to be a confluence of events—a mix of calamity, luck, and ambitious parenting—that has put him on the fast track to becoming, perhaps, the greatest runner in history. Certainly one of the greatest Europeans. You have to be made for that kind of spotlight. You have to be strong enough in the head not to let it get to your head. While Kristian Blummenfelt can wander around Bergen and go completely unnoticed, Jakob can't step foot outside his home in the Oslo suburbs without a camera on him, and it's been that way for quite some time.

Today, most of the attention and scrutiny he receives every day is his own doing, but for a long time, it was his father and brothers pointing the spotlight at the most talented runner in the family. By age 16, Jakob was one of the most well-known runners in Norway, and it was time to show off his talents to the world. All three brothers entered the mile at the Prefontaine Classic at the University of Oregon in 2017, where Jakob became the youngest person in history to break four minutes, doing so with relative ease in 3:58:07. Both brothers were five seconds ahead, and *Team Ingebrigtsen* had finally announced itself to the rest of the world.

Henrik, the most outgoing of the entire clan, wanted to capitalize immediately on their new fame—especially his little brother's—and began organizing a beer-mile (four laps of the track and four beers) world record attempt for Jakob. Given that he had just gotten his driver's license, enough reasonable adults squashed the attempt, much to the disappointment of Henrik and Jakob. They reverted to a more traditional means of accumulating fame and fortune: running very fast and sharing everything on YouTube. It's made them something of the first family of Norway, and it's had remarkable international appeal, which has helped secure lucrative sponsorship deals for the entire family.

A big component of the show's success is that it is so raw and almost voyeuristic in how close it gets to the Ingebrigtsens' day-to-day life. The "Kardashians of Scandinavia," as they are known, aren't scripting their reality show one bit. That would be way too stressful and energy sucking, especially for Jakob, who remains the most emotionally reserved on the show, despite coming out of his shell bit by bit over the years. Much of Jakob's emotional reservation is a result of his father's insistence. Like Olav reminding Kristian and Gustav that having a girlfriend would hinder their race times, Gjert declares girlfriends to be "the beginning of the end" in one episode. He was referring to the woman who eventually became Henrik's wife.

So at least for his younger years, Jakob believed emotions were something to be compartmentalized, shrouded and buried as deep as possible so as not to disturb training. It's quite possible that's changed since getting married himself in 2023 and welcoming a daughter a little more than a month before the Paris Olympics. Just don't expect him to start getting all emotional in interviews. "We're a competitive family, and so, in some ways, showing your feelings might be seen as a weakness," Jakob told the *New York Times* in 2022. "Feelings are not going to make you run fast. If you feel too much, then it's just an obstacle."

Controlling the narrative . . . and his coaching

Regardless of what a sports psychologist may have to say about a 21-year-old's views on feelings and running fast, it's hard to argue with the results. And Jakob has grown a lot as a human and runner in the few years since those remarks. He and his brothers cut ties with their father in 2022, citing physical and emotional violence that had resurfaced. It meant an end to the show—which all three brothers were happy to move on from—and coaching themselves for the first time in their lives. Jakob continued to be self-coached, even in the lead-up to the 2024 Olympics, with his older brothers acting as his advisers. Such an undertaking requires a certain degree of emotional maturity, even if he's not one to talk about emotions.

The kid who grew up very fast with others pulling the strings now has control over his own path and story and is better equipped to handle the highs and lows that he knows will come. And the higher he's climbed, the easier it is to find those lows. Much like Kristian, he reached the point where anything short of a win came as a disappointment quite early in his career. Second place at a world championship would've been a monumental achievement for 20-year-old Jakob, but by 22, it was a crushing defeat. That was yet again the case at the 2023 World Athletics

Championships in Budapest, where a violent kick from Britain's Josh Kerr in the 1500 left Jakob settling for second. He wore the emotions he doesn't like talking about plainly on his face in the finish photo that made the rounds on social media and particularly in news outlets in the UK. The cheeky Scotsman had finally gotten the best of the cocky Norwegian, and a budding rivalry was born one year out from the Olympics.

For Jakob, rivals are a bit like emotions. They're something he actively tries not to think about, and he rarely, if ever, speaks about them. In the aftermath of that defeat, the media wanted nothing more than to talk to Jakob about his new challenger, but he mostly kept his head down, claimed he hadn't been feeling well, and remained focused on the 5000 four days later. It didn't exactly go over well. Excuses seldom do, but claiming illness—however real or forged—is particularly frowned upon by athletes, the media, and message boards. It's seen as dismissive of the competition and even suggests that the athlete believes it takes some extrinsic force to be defeated. Which very well may be what Jakob believes.

Four days after what was probably the most emotionally crushing defeat of his career, he defended his world title in the 5000, holding off Spaniard Mohammed Katir, who had just stolen away Jakob's European record in the event. (Katir missed the Paris Olympics after whereabouts failures, or missing three drug tests, in 2023 and accepting a two-year ban.) Jakob has rivals coming from all directions—for once, there are men from his own continent and North America who can beat him in both events—just don't expect to hear much from Jakob about them, win or lose.

What we do know is that, despite winning his first gold medal in the 1500 and so many of his marquee moments coming from that event, Jakob's probably better suited to the 5000 and longer and has said as much. Even if he's also said that he doesn't relish going longer. There's a thrill in going fast, and there's a singular pain in going longer, as every distance runner knows. The 1500 is hard to win, and the races are often more tactical and explosive in nature, neither of which suits his strengths or method. That's part of the appeal. It's a race that amalgamates fast and long. Doing the 1500—or the mile—better than anyone before him is a good start to laying claim to the title of the greatest runner in history—a goal he mentions quite often. He'll likely have to do it outside of a championship setting, which rarely presents an ideal backdrop for Jakob's racing style. Without rabbits at the front to keep pace, Kerr's kick once again got the best of Jakob at the 2024 Olympics, as the Norwegian faded to fourth in the homestretch. A final burst from American Cole Hocker left Kerr to settle for second, but there was a certain victory in beating his new rival.

Once again, Jakob had to wait four days for redemption in the 5000 final, which he won by two seconds after stringing the 22-man field out like an accordion in the final 600 meters.

Jakob's perfect race

Jakob is not interested in winning a close race in the final 50 meters, and, as the world witnessed in his disappointing fourth-place finish in the 1500 at the Paris Olympics, he can't win that way against world-class competition. Jakob came with a lot to prove, despite—or perhaps because of—all that he'd proven already. He'd been marked for greatness since 2018, when he became the youngest-ever European champion at 1500 meters, but he was a grown man now, and if he couldn't grab the brass ring in this golden opportunity, when would he?

"Thirteen men in this field," said NBC commentator Tim Hutchings as they charged from the line on a sultry summer night in Japan's capital city, "and you can make an argument for every single one of them." While 8 of the 13 scrambled for pole position ahead of the first turn, Jakob elected to cruise in their wake, wait for the inevitable bogdown, and after two quick glances over his right shoulder to make sure the way was clear, swing wide and surge to the front. The pack strung out behind him, pressured by the Norwegian's forward-leaning, "I'm not here to fuck around" posture, which brought him through 400 meters with a spicy split of 56.14 seconds.

Also not fucking around was defending Olympic champion Timothy Cheruiyot of Kenya, who elbowed his way into the lead in the second lap, lifting the pace even further. Jakob couldn't have been happier. He had a hard race and someone else willing to share the work at the front. By 800 meters, which the new leader hit in a sizzling 1:51.76, the race was over for everyone except the defender and his challenger. But the challenger was on the rise, whereas the defender had had his day, and after biding his time till the final turn, the younger man returned Cheruiyot's earlier favor, shooting past him to gold and immortality.

Jakob wishes every race could be like that one. He wants to make the guys with more top-end speed uncomfortable as early as possible, like he tried and failed to do in Paris. It's similar to Kristian's ideal triathlon. He wants to go out hard from the start and then slowly turn up the tension until the rubber band is broken. Until the fittest athlete is pulling away from the ones with less fitness. It's rare that an athlete has a chance to celebrate the win before the finish in an Olympic 1500 final, but in Tokyo, Jakob had time to look back and verify that what he was seeing on the jumbotron was real—he finished nearly three seconds under the previous Olympic record despite slowing in the final 10 meters. He'd done to the 1500 what Usain Bolt had done to the 200. These were formerly events that were celebrated after the finish.

Maybe that's part of why he's come off as so arrogant to many. Perhaps it's the tattoos, which are a bit . . . different. His arms and legs are plastered with an assortment of icons and slogans with meaning and motivation to Jakob alone. Or maybe it could be his hair, which always seemed perfectly quaffed. It didn't help when he told reporters "You can't be disappointed with people not being better" after beating a world-class field in the mile at the Prefontaine Classic in 2022. He was confused by the follow-up questions and responded, "Everybody was running as fast as they could. They didn't choose not to follow the pace. They were hurting. And they're not better."

It's hard for that not to come across as domineering to his counterparts, whom he was calling anything but. He's been in front of a camera proclaiming that he wants to be better than everyone else since he was 10, and he's been better than everyone else at every age. But he's still so far from greatness, and that's why the matter-of-fact comments come off as so cavalier—for the time being.

Chasing down a legacy

Whatever medal count Jakob might achieve, until there are world records, it's unlikely he will be crowned the greatest of all time. Jakob may just be the best in the world at distance running's two shortest events for this stretch of history. Even still, he is perpetually talking about the work left to be done, and he appears to be closing the gap that remains between his running career and the great ones that have gone before him. He doesn't just need to break records; he needs to put them out of reach, like Moroccan Hicham El Guerrouj did with the 1500 and Ugandan Joshua Cheptegei did with the 5000. A record not held long is quickly forgotten.

But it's not about chasing any particular human from yesterday or today. To go down as the greatest to ever run, it would seem that the marathon would be an inevitable goal. But he hasn't expressed much interest in it thus far, and even giving it thought would take away from his current progression. One that he has complete control of for the first time. He knows he wants to be a complete runner, and for him, that means doing things like a cross-country season in the fall. So far he has seven European cross-country titles and lamented missing the event in 2023 due to a late-season injury. It's an important part of his development, and the vision is still long term. He's just not certain where that long-term vision will point next and how long he'll be able to keep building.

If he does want to retire as the greatest to ever run (for the time being), then at some point, he'll have to consider Kenya's Eliud Kipchoge, who is undoubtedly the best to ever do it on both the track and the road throughout a career. Of course, Jakob doesn't speak often of other athletes and history, but Kipchoge is assuredly on his radar. Alongside his brothers, he was the youngest pacemaker when Kipchoge unofficially broke two hours in the marathon in Vienna in 2019 at the INEOS 1:59 Challenge. That had to have left an impression on a 19-year-old about the weight of the marathon in the comprehensive chronology of running.

Wherever he decides to go next, he's running toward it with a foundation that few athletes with his talent have had before him and a unique support network in his two older brothers. They've taught him as much about what he should be doing as what he shouldn't, and it's the latter that may be his most peculiar gift. Whether it's from being pricked with a lactate meter for the first time at age 10 or watching his brothers get injured after trying to run immediately after long flights, he has a lifetime of learning about discipline and restraint. Another word for it could be suppression. He's mastered doing that when it comes to his emotions and his urge to run faster than everyone else. Save all of that for race day; otherwise, it's just going to cost you come race day.

BLUEPRINT

15

LACTATE TESTING: WHY THE TARGET IS LOWER AND SLOWER

A person should not agree today to what they'll regret tomorrow.

—BANDAMANNA SAGA

Once Marius Bakken began working with Peter Coe in his postcollege running career, he found that his body responded extremely well to higher-volume weeks, and that was the impetus for designing the first-ever training program built around double-threshold days. Depending on the time of the season, he performed either one or two double-threshold days each week, with anywhere from 20 to 40 percent of his total weekly volume done at threshold with a lactate monitor. That means he was doing up to 45 threshold miles per week, which would've been impossible for him (or any athlete) to do without controlling for lactate and/or breaking up the sessions into intervals.

Discipline was the key for Marius, and it remains the key for anyone attempting his form of training today. In order to get to the point where he could run 45 miles per week close to his own anaerobic threshold, Marius couldn't spend much time over it; otherwise, hitting his other goal of roughly 180K per week would become impossible. Trading a lot of volume for a little intensity was not something that had worked well for him, and it probably won't work for you if you've got big goals over long distances.

He also found—through a lot of trial and error, of course—that breaking the two threshold sessions into intervals ultimately allowed for more total mileage at or near threshold than was possible, with two continuous runs targeting a lactate level of 3.0 mmol/L for as long as possible. It's also the only way to measure lactate a handful of times throughout the workout and stay close to the target. Even within a given threshold workout, most of the intervals were performed just under threshold, with only the last one or two hitting 3.0 or perhaps a tick over.

Growing up outside Oslo, Marius was exposed to cross-country skiing at a young age and had a pretty good understanding of how the best winter sports athletes in Norway were training. He knew they were putting in more total hours of training at both high and low intensities relative to distance runners. He also understood that the mechanics of skiing and running are very different, and he'd been injured enough in his young career to know that simply making every workout longer would break his body to pieces. So instead of tinkering with the length of each session, he began experimenting with the frequency.

From a strictly mechanical perspective, training at the highest speeds possible can do more harm than good, and Marius saw that as he got older and learned more about his body and the bodies of those he trained with. He was mainly focused on the 5K and 10K distances, and running at threshold for 13–27 minutes requires a remarkable amount of metabolic fitness. At a purely cellular level, Marius didn't think training frequently at race pace could provide enough volume for the necessary changes to take place over the course of a year or many years. He had yet to go to medical school, but he'd learned enough about the cardiovascular system to understand that he'd be better off making his heart pump more blood than by making it pump more often. And his testing was beginning to show him that the way to do that was to do a lot more work at the lowest intensities.

After a performance plateau in the middle of his career, Marius found progress when he began running an average of 180 km (112 miles) per week, which was much higher than most of his European counterparts and more similar to the Kenyan runners who were so dominant at everything over 1500 meters. In order to sustain that kind of mileage, he had to remove any and all race-pace work because the muscular stress would become too much and the recovery windows too long. The ultimate goal for Marius Bakken—and the goal of anyone who seeks to fully embrace the Norwegian method—was to find that magical and often mythical sweet spot, which is as hard as one can push their hard sessions without inducing so much muscular and cellular damage that the next day or two are focused solely on recovery. That involves a lot of trial, error, and pricks for any athlete, but thankfully, there are a few decades of data to know roughly what we should be aiming for.

Why 4.0 makes a good GPA and an ambitious LT2

Way back in 1976, German scientist Alois Mader published a study that established 4.0 mmol/L as the anaerobic threshold in which sustained exercise was no longer possible in incremental lab tests. Millimoles per liter (mmol/L) refers to the concentration of lactate in a liter of blood. A result of 4.0 mmol/L is a pretty high number for most, and no one

Lactate Threshold Improves with Fitness

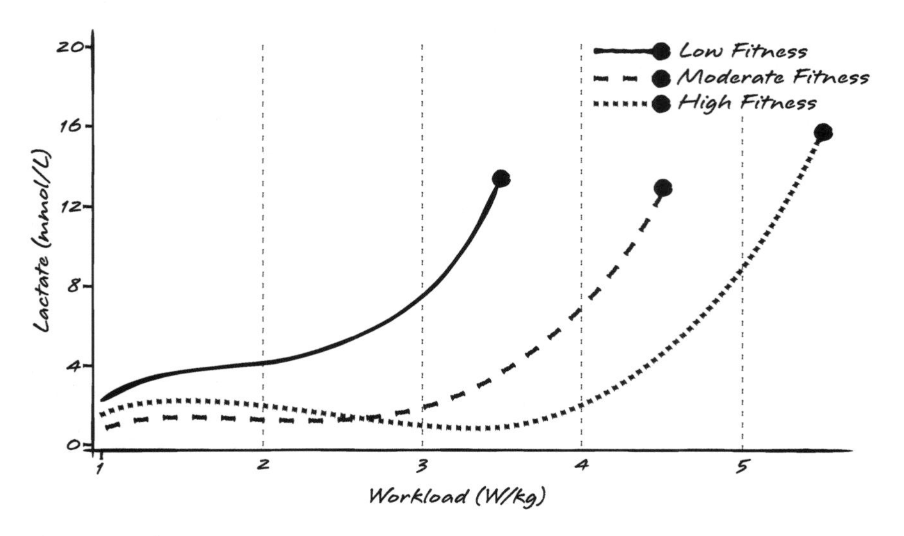

Fitness correlates with an athlete's capacity for greater workload, in this case quantified by power (watts per kilogram of body weight).

Source: Adapted from research by Jem Arnold.

is training to compete in an incremental lab setting. If you've ever had your blood lactate measured and seen a reading that high, you know you were probably going at a hard enough pace that you would have to back off pretty soon. For many, 4.0 is above LT2, or that second lactate inflection point, but there is huge variation between athletes.

When he first began using a lactate monitor daily, Marius found that he and his group performed most of their hard sessions at or near 4.0 mmol/L because the conventional wisdom of the day mandated that hard sessions should be nearly as hard as possible. It was only after training with a group in Kenya and performing lactate tests on them that Marius realized lower and slower was probably better for sessions most were doing higher and faster. He found that some of the best

Lactate Threshold Varies by Sport

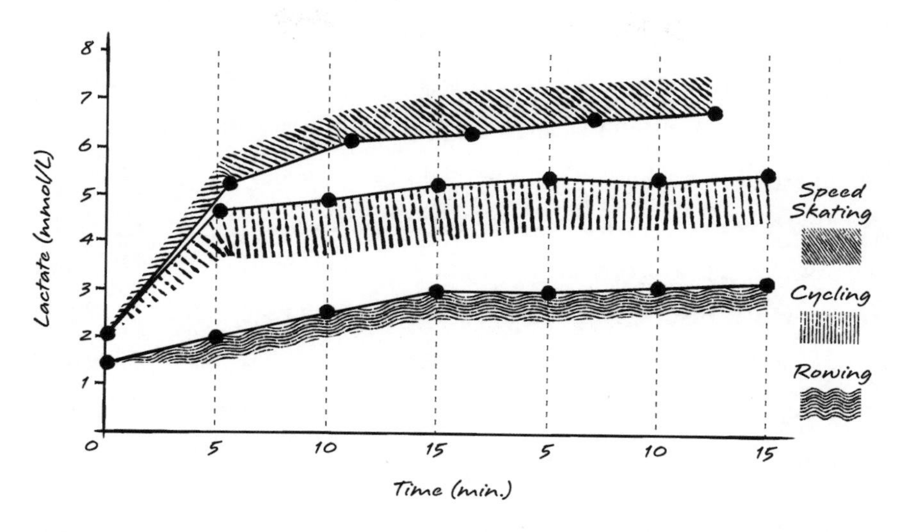

Blood lactate concentration in top-performance rowers, cyclists, and speed skaters during maximal lactate steady-state efforts show significant differences.

Source: Beneke, R., S. P. von Duvillard, 1996. "Determination of Maximal Lactate Steady State Response in Selected Sports Events." *Medicine & Science in Sports & Exercise* 28(2): 241–246.

Kenyan runners were doing their hardest sessions at lactate levels as low as 2.3 mmol/L. It confirmed his belief that what separated the very best distance runners from the next in class was that they were producing less lactate at higher intensities, not that they were somehow more comfortable at higher lactate concentrations. A couple trips to train with Kenyans also confirmed to Marius that the ideal amount of mileage was as much as an athlete could handle and that too much work above LT2 made it impossible to get the total volume high enough.

Marius adjusted his training to focus more on the lower end of his threshold, with race-pace work saved for race day or merely to test the results of the previous weeks. He and his cohort in Norway found at or around 3.0 mmol/L to be where the magic happened for them, allowing for an extraordinary amount of threshold work throughout the week coupled with plenty of low-intensity movement to create the necessary metabolic fitness. It was also controlled enough (i.e., slow enough) to allow for another critical element of the Norwegian method—the double-threshold day—which we'll soon explore in more detail.

While Marius and his training group from Norway are a very small sample size, the prevalence of lactate testing among both elite and amateur athletes over the past few years has demonstrated that 3.0 is probably a better place to start than 4.0—for most. If you take any size sample, it doesn't take long to see how much lactate and sweet spots vary widely between athletes.

Pinning down a highly variable target

Perhaps the only two athletes to have drawn out more of their own blood than Marius Bakken are Kristian Blummenfelt and Gustav Iden, and their own testing shows how individualized that sweet spot can be. According to Marius, "A major factor determining where an athlete's sweet spot lies is due to muscular factors, which in my opinion is the main limiter of training stress—and also the main reason athletes become overtrained."

Kristian, the more muscular of the two world champions, can perform the same threshold session as Gustav at roughly the same speed, much closer to 2.0 than 3.0, and can recover from those sessions faster. That allows him to do slightly more threshold and low-intensity training, peaking at nearly 40 hours per week. Both he and Gustav average roughly 30 hours per week throughout the year, and typically less than 10 of those hours are dedicated to "threshold" work with a lactate monitor.

Chances are your own sweet spot is much closer to 3.0 than 2.0 or 4.0, but the only way to know for sure is to purchase a good lactate monitor and get used to the half second of pain on your fingertip or earlobe. That's something the overwhelming majority of endurance athletes will find too disruptive to their training, and if the underlying goal is just to find fitness through fun, it may end up doing more harm than good.

While the lactate monitor is a core tool of the Norwegian method in the strictest sense, many of the training principles that have worked so well for the athletes who have embraced it can be incorporated into a less Norwegian program. It is possible to incorporate lactate-guided training into your own plan, whether you decide to bleed or not.

To bleed or not to bleed

We're all a bit trypanophobic. Some (like me) more so than others. No one likes to be pricked with needles, no matter how big or small, and getting comfortable with frequently jabbing yourself takes a small amount of courage and patience. It's also the single biggest barrier to entry for going all-in on the Norwegian method. Without involving a little blood, you're just doing a modified version of it. Which is fine, but probably not ideal from a strictly performance standpoint. For many of us, using a less invasive metric like heart rate or rate of perceived exertion (RPE) may be ideal from a joy standpoint. If pricking yourself a handful or a few dozen times a week makes you enjoy training less, it's probably not going to help you in the long term. But if you can learn to embrace it—or perhaps even enjoy the training more with the added layer of data and precision—it can unlock new potential.

Like any part of training, pricking your finger or ear during a training session gets a lot easier with time. It probably will never get to the point that you don't notice it at all, but it'll certainly be one of the less painful parts of your hard sessions. If you've made the decision to go all-in, the first question you likely have is which lactate meter to purchase. There are a growing number of (somewhat) affordable options on the market, but the two most popular have been around since the inception of the Norwegian method. The LactatePlus and Lactate Pro2 have been the gold standard for most of this century, but new models like the Lactate Scout+ and EDGE offer similar consistency and reliability. Like a power meter, the trick is to pick one and stick to it, especially when it comes to setting zones. The most affordable options are around $200, with the Lactate Pro2 costing twice as much. All the most current models can offer a reading in less than 15 seconds, which is key to preserving the integrity of an interval session.

The role of lactate in the athlete's body

Before we get into how to incorporate lactate-guided sessions into your training, a quick refresher. While lactate is often vilified as the limiter that keeps us from going harder for longer, it's just a by-product of our body's production of ATP (adenosine triphosphate), that magical nucleotide that turns calories into watts and, ultimately, velocity. During a long aerobic effort, ATP is produced via oxidative phosphorylation, which doesn't produce much lactate. However, it's important to remember that there's always lactate circulating in your bloodstream. Even the most well-trained athletes rarely dip below 1.0 mmol/L of lactate, and there's a good chance your own blood is concentrated with about 1.5 mmol/L even now. (Perhaps a bit higher if you're doing some easy training as you read this book.)

Lactate Curve from Lab Testing

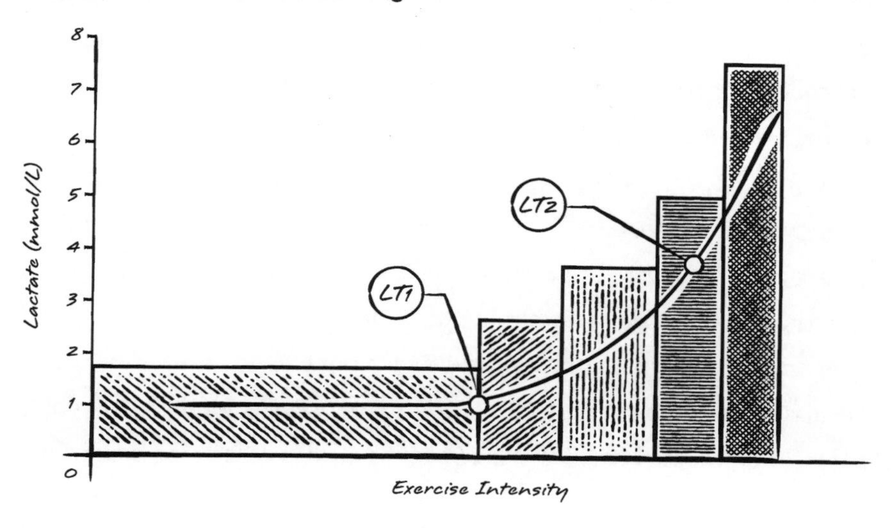

In a lab setting, training zones are derived from the key inflection points, which vary between athletes and within athletes over time.

Source: Adapted from Jamncik et al. 2020. "An Examination and Critique of Current Methods to Determine Exercise Intensity." *Sports Medicine* 50:1729–1756. https://doi.org/10.1007/s40279-020-01322-8.

At LT1, we start to see a slight rise in lactate concentration that is almost linear as the athlete moves from LT1 to LT2—let's call it roughly 2.0 mmol/L to 4.0 mmol/L to keep things simple. At this point, the body shifts to producing ATP via a process called glycolysis, of which lactate is one of the two primary by-products. The other is pyruvate or pyruvic acid, which is what lactate is eventually converted into within the muscle tissue thanks to a handy little enzyme called lactose dehydrogenase. This is important because pyruvate is a noncarbohydrate that can be converted into glucose when your muscles can't get any energy from oxidative phosphorylation (because you're working too hard). The more well trained an endurance athlete becomes, the more efficient their body becomes at this "recycling" of lactate, which is a crucial element in raising the anaerobic threshold. That's one thing the Norwegian method aims to be better at than other seemingly similar models—at least over the long haul.

Whether you're working off a three- or five-zone system, learning your two lactate thresholds is typically a more reliable way of finding your zones than less stable metrics like heart rate or RPE. You can do it without a portable lactate meter, but doing a lab test is often more expensive than getting one and doing it yourself. It's also a lot more accurate and is something every serious endurance athlete should consider doing at least twice a year to see if everything is going according to plan. One good race doesn't always mean things are on the right trajectory, just like one bad day doesn't mean it's time to start over. The lab rarely tells a lie.

In theory, the goal of any good training plan is to raise your lactate curve over the course of a season and many years so that both thresholds occur at a higher intensity. But it's not as simple as lifting one end of the curve to pull the whole thing up. A high tide does not necessarily raise both ships, and often one comes at the cost of the other. That's where individual goals—in terms of both races and what you're seeking to gain from lactate testing—become important.

The role of testing in your training program

The Bakken model of the Norwegian method uses the lactate meter to guide and control each session, most importantly the threshold work. That can work extremely well for certain athletes, especially those doing the most volume and demanding the most precision. But it's also quite intrusive. While Olav's approach still involves plenty of pricking, it uses lactate more to guide the coming weeks and months and ensure that there is positive progress taking place.

Tracking that progress doesn't have to be as complicated as plotting a lactate curve each week and seeing if it's moving in the right direction. After a season or two of testing, most athletes can see in real time if things are going according to plan, and time is a very important element in lactate testing. Adding time to a lactate reading all of a sudden makes it a two-dimensional point that is actually telling real information. The lactate meter only knows that an athlete's blood is currently reading 3.3 mmol/L. But if an athlete or coach knows that that number is coming on the fifth interval, whereas a month ago it was coming on the second, now they know something the meter doesn't, and they know they've made meaningful progress.

If you are interested in getting a lab lactate threshold test to determine your zones and thresholds, there's good news and bad news. The bad is that there probably isn't a sports science center close to you, and if there is, it's likely very expensive for an ordinary person to use. The good news is that many hospitals and cardiology centers can perform the test, and your doctor can write a prescription for one, which may be covered in part by insurance. If you can't find a lab near you that's available, you can do a do-it-yourself (DIY) threshold test with a portable lactate meter and a track, treadmill, or stationary bike.

A DIY lactate threshold test

The testing protocol is very similar to a VO_2 max test, with the main differences being that the intervals are slightly longer and you're not going to the point of complete exhaustion. Aside from a couple of pricks, it's a considerably less painful test and is repeatable with less cost—in terms of both money and recovery. The idea is to get three to four minutes at a given intensity to let the lactate build and stabilize, then take a reading and run the same interval at a slightly higher intensity. Here's what this would look like for a runner doing 800s on the track or treadmill:

- » Take baseline reading.
- » Warm up with 1200, very easy.
- » Take a second baseline reading.
- » Run 800, easy.
- » Take a third lactate test.
- » Run 800, moderate.
- » Take a fourth lactate test.
- » Run 800, hard.
- » Take a fifth lactate test.
- » Run 1200, very easy.
- » Test a final time to see if you've returned to baseline.

Every lactate meter will come with software that will plot your results, where you'll be able to see at what pace, power, and heart rate each rise in lactate occurs, similar to the previous graph. Whether you plan on making lactate testing a regular part of your training or not, establishing those two data points can provide the guardrails that are fundamental to the Norwegian method of training:

1. Knowing LT2 will ensure your threshold work doesn't get too hard.
2. Knowing LT1 will ensure your low-intensity work is easy enough to maximize the volume of low-intensity work and subsequent threshold days.

If you don't plan on using a portable lactate meter to control threshold sessions, heart rate is the next best bet for runners or power for cyclists, but know that whatever your metric, it will vary with factors like fatigue, heat, and altitude.

The more often you test, the more you learn about how lactate varies throughout each workout, which is really telling you what energy systems are at play. When athletes are just starting out, especially if they haven't warmed up, most of the energy comes from the glycolytic system, causing a slight increase in lactate values at the start of the workout. Don't be surprised if the first reading you ever do on yourself is much higher than expected during an easy effort. It doesn't mean you're out of shape; it means you need to warm up so that your body can supply more energy through oxidative phosphorylation. Essentially, the effort becomes more aerobic and the values come down and stabilize until you do something to disrupt it.

As you test more often, you'll also see that dehydration plays a huge role in lactate readings. After all, the meter is only able to measure the concentration of lactate in a fixed amount of blood. If your total blood volume is reduced via dehydration, your readings will go up, and it can be quite substantial. It doesn't mean that you're producing more lactate, just that there's the same amount in a smaller container.

Think of it like a cup that's half honey and half water. Leave it out in the sun for a day, and a good bit of the water will evaporate, but the honey will remain the same. Now you have a significantly higher concentration of honey without producing more. This is what's happening to your lactate readings at the end of very long sessions or races, so don't be alarmed if they occasionally pop off the curve (especially for Ironman athletes in hot climates).

Even if you're not testing regularly, or you're not adhering to the Norwegian method in any of its manifestations, a regular lactate test a few times a year can be vital for you and your coach, and it's not very disruptive to the overall plan. Typically one is done during the offseason and then after each big training block during the year. It's one of the best tools that coaches have to test the results of a training or altitude camp without throwing an athlete right into a race. It's not a test that an athlete should do completely fatigued, but it's also not one that requires any special rest or recovery. Do not taper for a lactate test or even a VO_2 max test, which you should know is much more painful, even for trypanophobic athletes.

Your own world-class coaching consult

While the Norwegian model is lactate centric, like any training model, it's important not to get too caught up in lactate whether you're trying to make your own training model more Norwegian or just more precise.

Olav is often quick to admit that lactate gets too much attention when it comes to his own model or the Norwegian method in general. He knows the history of the method, but he prefers to use lactate as a frequent guide to see where the program is going so he can make tiny changes along the way instead of big changes at turning points, and that's how he believes most age-groupers can benefit from it as well—whether they choose to test a few times a year or every week:

> Everyone is getting too caught up in curves and inflection points. But I try to remind people that each point on the lactate curve is one-dimensional. For every point on the curve, you have reduced a single information point: At 370 watts, I'm at 2.1 mmol/L. But

now what? Then people say, "We need to find the inflection point," but we think that's very important for some athletes and not very important for others in terms of where they need to be focusing their training.

Say a world-class coach has been coaching athletes for years and it's going well and they've been using 4.0 mmol/L as a standard for threshold training. Now I come along and say, "No, no, your threshold should be LT2 for each athlete, and you have to find that inflection point for each of them." If they all of a sudden incorporate that without understanding the implication on the whole program, you can end up reducing the total stimulus on your athletes. The reason the athletes were able to get as good as they are in the first place is because the coach found a balance between intensity and duration that works for them.

When you communicate to your athletes, this is a threshold session, they have an understanding of approximately where they should be and what's worked for them. If you suddenly have this epiphany because you learned about inflection models or some other definition of threshold, then you end up bringing the total training stimulus down. Maybe they're now doing those threshold workouts at 3.0 instead of 4.0 because that's the inflection point. And maybe they do get faster over the short term because it's almost like a minitaper. But over time, the total volume and stimulus are reduced.

The substantive takeaway for coaches and athletes is that if your program has been working well enough that you're considering introducing lactate-guided training or regular lactate testing, don't overhaul everything all at once if your numbers aren't what you expect or they're different from what's worked for others. It can be a powerful guiding tool in making your training more precise and more adaptive, but—unless someone is starting this kind of training as young as the Norwegian champions we've met—it's best treated as a cherry on top and not the ice cream.

BLUEPRINT

16

THE BIG THING: DOUBLE-THRESHOLD TRAINING

Only idlers wait till evening.

—LAXDAELA SAGA

No single component of the Norwegian method has garnered as much buzz as double-threshold sessions, and for good reason. It's perhaps the simplest aspect to implement into one's own training; the results at the elite level are getting hard to argue against, and it's pretty difficult. Whether amateur or pro, those competing at the pointy end of the races are typically the ones most turned on by doing very hard things in training. You're not reading this book if you don't enjoy putting yourself in the hurt locker from time to time.

A double-threshold day is precisely what the name implies: two threshold workouts done on the same day, typically with three to eight hours in between, and ideally guided by a lactate monitor. The reasoning for doing multiple hard sessions in one day instead of spreading them out throughout the week is similar to the reasoning for controlling lactate: it requires less recovery, it allows for more overall training volume, and it leads to improved metabolic fitness and higher thresholds over time.

A melting pot of training methods

Of course Marius Bakken wasn't the very first athlete or coach in history to mess around with multiple workouts in a day. Anyone who grew up competing on a swim team probably feels like they were doing some version of double-threshold training by the time they were 12. When it comes to running, renowned 1980s Italian coach Renato Canova used what he called "special blocks," which was a week or two of very hard and high mileage ahead of big meets, and this frequently included multiple hard sessions a day, done multiple times per week. Even before he began working with Marius, Coe would occasionally prescribe multiple hard sessions in a single day during what he called "top-up" weeks. These were simply weeks in which Coe ordered his athletes to do as much mileage as they could withstand to ensure aerobic endurance wouldn't be a limiter. One could describe the Norwegian method as polarized training layered with Canova's special blocks and Coe's top-up weeks.

That's also where Marius's Norwegian method 1.0 (or 2.0, according to Dr. Seiler) differs quite a bit from the versions used by the great Norwegian athletes of today (or non-Norwegians practicing some form of it today). Marius was training for something very specific: the 5000 and 10,000 on the track. He was building a model to excel at exactly that, which is why his trial and error brought him to that magic 3.0 number for his specific body using his specific lactate meter. Thresholds, meters, distances, and goals will vary from one athlete to the next, and any good training model needs to be adjusted accordingly.

The double-threshold days Marius was performing looked very different from those of Jakob, Kristian, or Gustav because they're all training for very different races, and they're very different athletes. Because so much of the understanding of the Norwegian method has come from Marius's own writing, some assume that all practitioners are hovering around the magic 3.0 number for all their threshold workouts. This, of course, isn't the case. For races as short as a 1500 on the track or as long as an Olympic-distance triathlon, sprinting at the absolute limits of lactate concentration is often necessary to beat the best in the world. That needs to be trained for—in a very limited manner—and usually as close as possible to when you'll need it. In that sense, the training of a Jakob or Kristian isn't exactly the most relatable for age-group athletes who are Norwegian-method curious.

How to Build a Base that Supports Double-Threshold Workouts

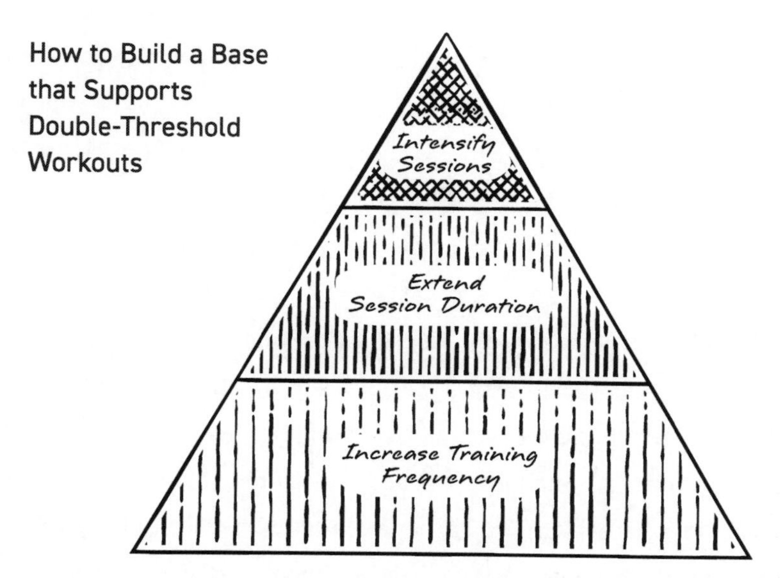

Before adding intensity, build volume—first in the number of days you train, and then in extended sessions. With an adequate base, the high-quality sessions at intensity will be more rewarding.

Amateur runners need to take particular caution if attempting double-threshold days because most haven't—and probably shouldn't—attempt double runs of any form because they lack the foundation. In Marius's personal experiment, he introduced the double-threshold days mostly as a way to get his mileage up beyond 100 miles a week. He and many world-class runners could get close to 100 without the need to double down on hard sessions or even double at all. So if you're considering double-run days and have been doing 40 or 50 miles a week, there's probably more cost than potential benefit.

For Marius, after just one season of double-threshold training, the results were more profound than he could've hoped. Within a year of embarking on his homemade program (which included plenty of Coe's influence), he dropped his 3000-meter PR from 8:13 to 7:47 and set a Norwegian record of 13:06 for the 5000 in 2004. That record would stand for 15 years, until it was shattered to pieces by Jakob, who was doing the kind of training as a teenager that Marius was doing in his mid-20s.

As is often the case in high-performance coaching, Jakob's longtime, albeit former coach has brushed aside Marius's influence in the development of his son's program, but the influence is in plain sight. Literally. After Gjert made a claim that he had no knowledge of Marius's training protocols, Marius published an email communication on his blog proving that the Ingebrightsen patriarch had requested and received the details. It's clear that all three Ingebrigtsen boys benefitted from the training principles that began with Marius. Or Coe. Or Canova. Or maybe it was Ingrid. It depends on whoever is giving the credit, but it certainly didn't begin with the Ingebrigtsens.

Sebastian Coe, who followed up a decorated running career with a tenure as president of the International Olympic Committee, summed it up well when asked about his father's influence on the Ingebrigtsens' training program in a 2023 interview for LetsRun.com: "I am a big believer of understanding the history of the sport. . . . Every generation is influenced by groundbreaking thinking of the previous generation. Become students of your sport. Good coaches do that."

Double threshold to the power of three

When it came to adapting the Norwegian method and double-threshold training to tri, credit goes to Arild, and it's become a laser-focus for Olav as he's taken over the reins of coaching Norway's and the world's most decorated triathletes. One thing that's become quite clear as Kristian and Gustav piled up world titles is that the method largely cultivated by Marius for running on the track may be even better suited for triathlon, where the shortest races are roughly the duration of a marathon and the longest take most of the day.

The Norwegian method of running and triathlon are as different as the sports themselves, and many believe the latter is a bit behind even still. Olav doesn't necessarily buy into the idea that the method is better suited for long-distance triathlon than short. He identified a sizable performance gap between Olympic-distance and Ironman training, and that's why they've been able to lob chunks off records that had stood still for decades. As he saw it, the top performers at the Olympics were much closer to achieving absolute peak human performance than those competing at Ironman, given that Olympic-distance races are most often determined by the best 10K run and humans have gotten pretty close to their peak at running 10K. Running a fast marathon on tired legs after a 112-mile ride and a long swim is something humans are just beginning to figure out. For the most part, the

A High-Intensity Interval Pop Quiz

A constant workload of high-intensity exercise starting from rest will likely resemble which of these blood lactate curves?

Exponential

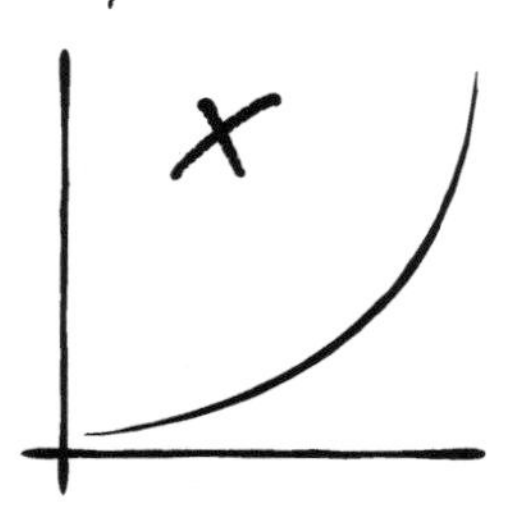

This is what a lactate threshold test feels like—it will push you to your breaking point, but this is not the goal of an interval workout.

Linear

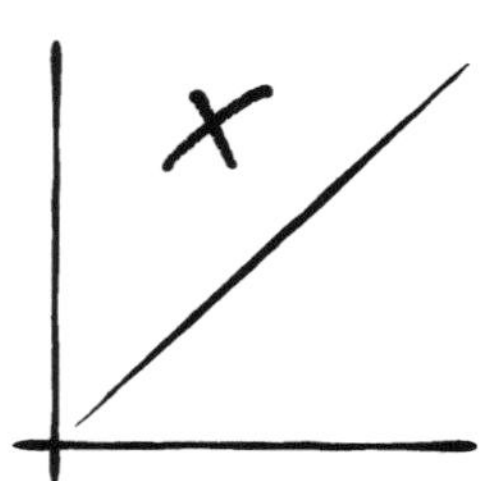

This is the expectation athletes bring to training—that input is rewarded in equal measure, a highly unlikely outcome.

Logarithmic

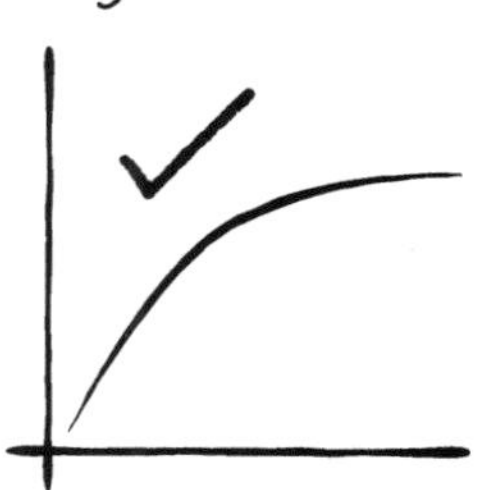

Once you know your lactate thresholds, the goal is to stay in your zone for a given set of intervals—don't ruin tomorrow's workout.

longer the distance, the less we know about training for it. Perhaps it's oversimplifying the problem, but in Olav's own words, "With more volume has to come more precision in everything you do; otherwise, it will tip the load over."

The double-threshold training of Marius and Jakob involves two running sessions on the same day, usually on the track, but triathlon offers far more possibilities. (The treadmill can be a useful tool for runners, particularly in northern latitudes and any time heat training is in order.) The Norwegian triathletes—whether Kristian and Gustav training with Olav or those training with Arild on the national team—almost never perform two running sessions on the same day and rarely two threshold sessions of the same discipline on the same day. Kristian and Gustav have traditionally done two double-threshold days per week—usually Tuesdays and Thursdays—but the mix is either swim/bike, swim/run, or bike/run. The order of the disciplines doesn't matter for the sake of training stimulus (i.e., there's no difference between bike/run or run/bike). It's merely a matter of athlete preference and how the coach is integrating the workouts into the broader plan. If anything, a run/bike double threshold can be more beneficial for the workouts in the days that follow because the sport that requires more mechanical and gravitational stress is first, allowing for more time between the intensity of that session and the next day's work. But whatever the double-threshold day holds, it's critical that the coach or athlete zoom out and organize training in a manner that works within the overall plan.

Dialing the dose-response equation for triathlon

Even when Kristian and Gustav are training for an Ironman, they don't do many steady-state efforts, a staple in the training regimen of long-course triathletes. They will have rides and runs that are close to five and three hours, respectively, but within that time there are plenty of intervals and efforts built in. In Olav's mind, a steady-state session simply can't pack the same amount of stimulus, which is a waste of time or opportunity. It's also why he believes the double-threshold model is so effective: It allows you to dose the load out across two workouts for more total stimulus. He looks at it as breaking up the volume of high-intensity work over the duration of a single workout: "Let's say that you're able to go 400 watts for 5 minutes, but you're completely washed after that effort. But if you take those 400 watts and now hold them for a 1- or 2-minute interval, then you will be able to do quite a bit more work during one session. Maybe you're able to do 8 or 10 minutes total at 400 watts in 60 minutes instead of 5 [minutes]. I think we'd all agree that that's quite a bit more stimulus."

Stimulus drives adaptation, so in the end, the goal is to increase the stimulus and drive meaningful growth. And the need for stimulus is matched with a need for volume. As the Norwegian method sees it, more volume will almost always lead to more success as long as you're managing the balance well. It's a tenant that crosses the boundaries of both discipline and distance, applicable to both sprinting and endurance sports. Olav doubles down on the importance of building volume, even for sprinting: "As long as you're able to manage the load, more stimulus will always be better."

Going to complete exhaustion leaves little room for volume on subsequent days, and there's not that much stimulus that can be gained from a single, big session. More important is how each session of more minor stimulus is pieced together over the long term.

As Olav zooms out to the training week, there's a grand scheme at work of high-intensity and low-intensity workouts that function as intervals and rest. Even the week of training reads like an interval session. With this perspective, he looks for opportunities to take one day's session and spread it out over two, creating a higher, more sustainable stimulus in the long run—with precision comes growth.

For Olav, the magic of the double threshold is that there is none. It simply has to do with dialing the dose-response equation, and he thinks double threshold offers a smarter way to get the largest dose possible to create the biggest response. And like Marius, it's taken a lot of trial and error to find a model that makes for the best long-term aerobic and anaerobic growth. It's not about a specific modality or workout, and even Olav rarely uses the term *double threshold*. He also much prefers dose and stimulus, and for the time being, two double-threshold days per week seem to be the best way to get it done. Although he'd be the first to admit that maybe there are better ways to get an even greater total stimulus and hopes to find those in the future.

Even control has its limits

If you were to hire Olav to coach you (which you cannot), he would be quick to tell you that you don't need a lactate meter to do your own double-threshold training. Whereas Marius and, to a large extent, Jakob use lactate to control and monitor a given session, Olav's model uses lactate only to gauge progress and adjust the plan as needed. The "threshold" sessions, particularly when the athletes are training for an Olympic-distance race, are much less controlled than anything Marius would've done over his career.

This could be largely a result of the different sports. When you consider that swimming and biking have considerably less mechanical cost, injury risk, and gravitational stress than running, it stands to reason that you can get away with more. Triathletes are also training for something more nuanced than holding the best pace they can for 5000 or 10,000 meters. Consequently, the lactate meter is on hand for threshold work, but it's not like a little spike is going to change the plan for the day's workout. The goal is to collect data for the future, and it's the long game that Olav feels will reveal the real power of the Norwegian dedication to lactate testing: It allows them to track progress and be more fluid with the plan from week to week. The only better way to do it would be to race nearly every week, which isn't realistic. (Although they do race a lot and are almost always race ready, which is something we'll get into in Chapter 17.)

As far as organizing the double-threshold workouts themselves, Olav's model uses mostly pace and power, which is good news for anyone trying to replicate it at home with no interest in drawing blood. If you have Olympic medal aspirations, doing double-threshold workouts without a lactate meter may put you out of medal contention. But barring such aspirations, pace and power will work just fine. (For triathletes who are considering the addition of either a power meter or a lactate meter, the power meter should come first. If a coach tells you otherwise, it's time to find a new coach.)

Build volume before you double down

High-mileage runners and triathletes with a lot of miles in their legs can benefit from incorporating double-threshold days into their plan, especially if they've been training for a few years (or decades) and are accustomed to working out every day of the week. After all, more stimulus is always going to require more time. Let's say, for example, that you're used to doing two hard 60-minute sessions per week that bring you close to exhaustion. It would be ill-advised to add another hard session to one of those days. But if you take that 60-minute session and shorten it to 40—and layer enough control that you're not tipping over to exhaustion—then it's probably possible to perform another session of roughly the same time and intensity on the same day. Much like how one steady state will bring you to exhaustion faster than if you break it up into intervals, if you break that hard session up into two intervals, the total stimulus is going to be more, and the level of fatigue will be the same—or even less—for the following day. Adding that up month after month is where the magic is made.

Even if you don't have the time, need, or desire to incorporate true double-threshold training into your plan—or if you're like Olav and you prefer to stay away from the term *threshold* altogether—splitting "normal" sessions may have benefits for the short and long terms. It's generally going to mean a higher volume throughout the week or month, and not only will the kick in metabolism help certain athletes achieve a higher portion of lean muscle mass over the short term, but it'll also help all athletes become more metabolically efficient over the long term.

It's something nearly all elite swimmers have been doing for a century, and the reason that is often cited is that the near absence of gravity in the water makes it possible to do a lot more volume (in terms of time) than running. That's obviously true, but it's also the reason that long-course triathletes from a swimming background have had a remarkable amount of success. From Mark Allen to Lucy Charles-Barclay to Kristian Blummenfelt, the ones who grew up swimming are rightly credited with having a massive aerobic engine from a very young age.

While some pockets of runners have heralded "two is better than one" going back almost a century, it's fallen mostly on deaf ears. In recent years, Flagstaff, Arizona, has become something of an American hub of the Norwegian method—and fittingly so since it's also become Jakob's favorite place to get in his altitude. Even after winning back-to-back National Collegiate Athletics Association (NCAA) cross-country titles, Northern Arizona University (NAU) head coach Mike Smith introduced double-threshold training for his athletes in 2019 as a means of increasing volume without increasing stress and fatigue. It's part of what has helped a relatively small school stay atop of NCAA ranks, and it's spread throughout Flagstaff. It's become a staple of the marathon-focused athletes on the HOKA NAZ Elite team as well as frequent Flagstaff visitor Hobbs Kessler, who finished fifth in the 1500 in Paris at age 21.

The one thing all these athletes have in common is that they're running very high mileage. Even the youngest runners on NAU's teams are doing 70–80 miles a week before they get there. The medals and world titles won by athletes doing double-threshold training have thrust it into the limelight like never before and may have more amateurs considering it than probably need to. But if you're already up in the stratosphere where real runners would deem high mileage, and you've hit a plateau, the road back up the mountain may be splitting up some—or all—of your hard sessions into two.

For mere mortals, the risk of trying double-threshold days may exceed any potential reward. And if you're already nailing a couple of hard sessions a week and getting in plenty of volume around those, you're probably moving up the mountain just fine. Breaking up those hard workouts into two can run the risk of lowering the intensity and spending too much time in the gray zone that leads to a plateau. And no one can reach their peak from a plateau.

BLUEPRINT

17

THE ART AND SCIENCE OF PEAKING

Many are wise after the event.

—FLJÓTSDÆLA SAGA

Shortly after obliterating the two-mile world record at the Paris Diamond League meet in June 2023, Jakob Ingebrigtsen told *Citius Mag*, "One of the biggest mistakes that a lot of people do is they go too hard in training." He went on to say that this mistake is a telltale sign of a lack of belief. When an athlete is struggling with the idea of whether or not they can do something, they try to do it repeatedly in training. Instead of banking evidence of what they can do in training and building confidence, they end up eroding their readiness on race day.

His comment was mostly passed around as arrogant on social media—like Jakob was yet again calling out his competition—but he was highlighting a core tenant of the method that he's lived by. Ask a dozen of the most prominent coaches and athletes using the Norwegian method to describe it, and there will be a lot of variation and nuance. Ask them to describe the biggest mistake other athletes are making, and their answers are likely to be more similar, something like "They're going too hard too often in training."

That was the point Jakob was trying to make. He never goes harder in training than he does in a race—whether that's according to RPE, heart rate, or any other metric imaginable. In the same interview, he was asked if it was his training or his mental toughness that had him so far ahead of the rest of the world at any distance from 1500 to 5000. He responded that it's probably neither and likened race preparation to building a house. As he sees it, his competitors are great at building the foundation and walls, but where they go wrong is during those final five or six weeks before a big race. They put on the roof incorrectly, and the entire house comes crashing down. He's not just talking about a blown taper, which is an excuse both elite and amateur athletes like to throw around after a race doesn't go according to plan. Jakob is talking about blowing the most important weeks of training by feeling the need to prematurely test all that fitness that was built in the months and years leading up to it.

Saying your competition lacks confidence is a surefire way to get labeled as arrogant. But it's not much different from an athlete saying they've outworked their competition, that they've gone deeper into the well, and that their hardest sessions were the hardest of anyone on the starting line. For a long time—and still to this day—many athletes feel the difference in reaching a new level lies in continually going deeper in training: the "do more harder" approach that held back Mark Allen at the beginning of his career.

If Jakob's post-race comments sometimes come off as arrogant, then Kristian and Gustav's pre-race practices might qualify as a new level of cavalier. The notion that the controlled method of Norwegian triathlon training is somewhat easier than those that have preceded it was discarded when the two Norwegians completed more than an Ironman over the course of two days just one week out from the 2022 Ironman World Championship in Kona. While most of their competition was spent greatly reducing volume and sprinkling in a little extra speed work along Ali'i Drive in Kailua-Kona, Hawaii, the pre-race favorites were busy with a massive two-day effort. They did a race-pace swim followed by riding the entire bike course on Saturday. The ride was relatively steady state by their standards, but there were intervals to hit race pace and breaks for lactate tests. The following day they ran 40K at race pace—a little more than a mile short of a marathon—again with just a few short breaks for testing and guidance from Olav. Even before race day, it caught wind on social media and was the talk of the tiny island town. Surely the Norwegians had blown their taper with such a massive effort so close to such a hard race.

But it worked. Mostly because it was still a reduction in volume from what they would typically do over a weekend, and it was certainly easier than what they were planning on doing come race day. It worked especially well for Gustav, who won and set a new course record in his first race in Hawaii and only his second-ever Ironman. His time of 7:40:23 was almost exactly what he and Olav had aimed for. They had expected to finish one minute faster, based in large part on the data from the big workouts leading up to race day—like the back-to-back days that had caught so much attention. Kristian and Olav had also reverse-engineered a plan for a 7:39 finish, but unfortunately for the then reigning Ironman world champ, his legs couldn't respond to Gustav's kick late in the marathon, and he finished four minutes behind, in third place. It's the kind of result that would typically send

Kristian spiraling into despair, but he handles losing much better when Gustav comes out on top, because he had a hand in the victory, and it's a win for the team of three.

You can blame swimming for your ineffective taper

Just as there is a wide range of definitions around thresholds and zones, the endurance world has very different ideas on how to execute a taper. Some look at the word as simply meaning a reduction in training and adjust their volume accordingly. But outside of sports, it is defined as "narrowing toward one end," which might also apply to the context of preparing for an endurance event. The athlete is simply narrowing their focus to the specific demands of the race while discarding all unspecific work, effectively reducing volume and optimizing recovery.

In the world of sports, the taper terminology first caught on among swim coaches in the 1960s, but of course what works in swimming can spell disaster in true endurance events. Both an elite marathon runner and an elite 100-meter swimmer can be training 20 hours per week. One is tapering for an event that might last 2 hours and 15 minutes, while the other is tapering for what will probably be a 45-second race. The marathon runner is aiming to run roughly 11 percent of their weekly training volume on race day. In terms of time, the race itself isn't far from what the athlete would be doing on a typical big day. The swimmer, however, is tapering for an event that is 0.065 percent of their weekly volume. They're getting ready to race a distance that wouldn't even account for 10 percent of their warm-up.

If the goal of a taper is to minimize fitness loss while maximizing form (i.e., sharpness), then the marathon runner has a lot more to lose than the swimmer. Aerobic fitness is still paramount when you're running 11 percent of your typical weekly volume in a race, so a massive reduction in volume doesn't make much sense. Form matters quite a bit more when you're getting ready for a single anaerobic effort that isn't even 1 percent of your typical daily volume, so a big reduction in volume makes sense. It's easy to see that the taper for more anaerobic events and endurance events is very different, and we might even agree that it merits separate terminology. It's taken a few decades for the endurance taper to distinguish itself from the swim taper, so maybe the vocabulary will follow suit.

For swimmers, in the final three weeks before their "A" meet, the goal is to let the body make the final adaptations from months of overload, at least relative to the races they're preparing for. And the balancing act between volume and intensity isn't nearly as delicate as those tapering for true endurance events, whether it's 3 minutes or 24 hours.

Dialing volume and specificity for endurance racing

For the endurance athlete, the reduction in weekly volume during the final three weeks before an "A" race should be much less dramatic than the swimmers who may have been first to apply the word to their sport but certainly didn't make a one-size-fits-all-sports plan. And it also should look much different from Kristian and Gustav, whose typical weekly volume is so high that even with a slight reduction, it still looks like a massive load if you only look at the day or week.

Let's say you're training roughly 12 hours per week and hoping to pull off a 9-hour Ironman and Kona slot. It's ambitious, but people have done it in considerably fewer hours of training. Race day is going to encompass 75 percent of your normal weekly volume—all done at whatever you feel is your threshold, ideally, and that's going to require special recovery beforehand. That has to come at the expense of volume because, if anything, intensity—or better yet, race-specific work—needs to increase or at least stay consistent.

Obviously marathon runners are completing significantly less overall volume on race day relative to Ironman athletes, but the overall training volume is often reduced too. (It helps when you don't "need" a five-hour ride almost every weekend.) To keep things simple, say a runner is training nine hours a week and hoping for a three-hour marathon to qualify for Boston. Race day is still 33 percent of that weekly load. It may not sound like much compared to an age-group Ironman athlete, but it's still a huge day that needs special concessions. Just not too many and not that special.

What you can learn from Jakob's taper

Compared to Norway's most famous triathletes, the taper for its best runner looks relatively easy. But the principles are the same: Don't change much, and keep things as specific as possible. And by all means, save the biggest effort for the race.

While Kristian and Gustav complete their final key sessions very near race pace (Olav estimates that they're often above 95 percent of race pace or power), Jakob proudly claims that he never goes above 87 percent in training, even during his final weeks before a major meet, when he often has a race or two in which he goes very close to 100. Surely he has heaps of data to home in on that 87 percent number, but the fact that it's such a specific number is not something to obsess over and is still somewhat of an estimate. It really means he feels he's at an 87 RPE if the scale goes to 100.

Obviously 87 percent of Jakob's max speed for 1500 or 5000 is still faster than most humans can run for 200 meters. It's very fast running, but the confidence comes in knowing that he could still go harder—and that it will be there when he needs it. And there needs to be some concessions from his typical routine of between 110 and 120 miles per week spread out over as many as 12 sessions. Luckily for those of us trying to glean something from the effectiveness of Jakob's taper, like his triathlon countrymen, he's been very open with sharing exactly what he's doing.

Similar to the training plan Olav has designed for Kristian and Gustav, the plan originally formulated by Gjert Ingebrigtsen is very similar from week to week and was designed, in part, to keep his sons in race shape nearly all year round. One of the key tenets of the Norwegian method in all its forms is to race often and always be race ready. If you're not going to test your absolute limits in training—as Jakob is so fond of reminding he does not—you still have to figure out where they are by racing against worthy competition with some frequency. And with all its layers of control, the Norwegian method should, in theory, make for more injury-free seasons with uninterrupted race schedules. (Save for Achilles issues for Jakob and Gustav in recent seasons and a stress fracture for Kristian when he was still a teenager, the three Norwegian superstars have largely avoided injuries throughout their professional careers, which began at age 16 for all three.)

Jakob's workout that has received the most attention over the past few seasons is his infamous hill session, which typically comes on a Saturday—one of the only days he won't do a double workout. It's two sets of 10 reps of a 200-meter hill by his home, all done above threshold, again, somewhere around 87 percent of his max effort. It's an incredibly hard workout when done at the proper intensity, and he's been adamant that it is by far the hardest session of his week. It's also the one workout that disappears on race week because it's not specific enough, and it demands so much recovery that even he can't back it up with a second workout on the day (most of the time).

Other than skipping his hill session, the plan mostly stays the same until three days out from a race. Jakob is still running twice a day, just trimming off a bit of volume throughout the week and making his final intense workouts—two days before a race—as specific as possible. A few of his favorite key sessions two days out from race day are five sets of 300 at 1500 pace or five sets of 5000 at 5K pace. The day before a race is the only time Jakob truly takes it easy, with just a couple of minutes of intervals at race pace mixed in with some easy jogging. He's noted that he does a maximum of four hard minutes the day before a race, whether he's getting ready for the 1500 or the 5000.

The magic of the method

As with most inventions of the Norwegian method and its interpreters, the magic of Olav's mythical taper for Kristian and Gustav leading into Kona 2022 was that there was none. It was just a narrowing of focus—making the hard sessions as specific to the demands of the race as possible and doing so with a slight reduction in weekly volume to give the muscles and soft tissue a jolt of accelerated reclamation. When your body becomes accustomed to 30-plus hours of training per week, a few 25-hour weeks can have the muscles fully recovered with little loss to the aerobic engine. And that's really the juice of the method and the reason athletes like Kristian, Gustav, Jakob, and Ingrid in her heyday can do seemingly superhuman things in the days and weeks leading up to superhuman performances. It's because they make the seemingly superhuman ordinary, and they've been doing so for a very long time.

Chances are you haven't been logging 25 hours of training per week since you were a teenager and that your annual training plan looks a bit more pyramidal than Norwegian—for now. Perhaps that'll change. Most of us don't enjoy the lifestyle of a professional athlete—Norwegian or otherwise—and we can't layer month upon month of consistency the way these elite athletes do. Your training plan (and taper) will require a few more concessions and certainly less volume. The ticket is to make every workout count—whether it's an early season long run or race-specific threshold session—and take into account the demands of what lies ahead. At the core of the Norwegian method is control. It's especially paramount the closer you get to your long-term goal, whether that's a few months or a few years down the road.

EPILOGUE

It was appropriately pouring during my final hour in Bergen. The only time I'd seen the sun over the past three days was when it was keeping me awake at 10 p.m. Kristian was enjoying a few pastries and coffees while I finished a long interview at a café that I was very familiar with by then. The man can consume carbohydrates and caffeine like no other human I've met, and I go to extremes with both. We've talked about training philosophies and techniques—particularly when it comes to swimming—which was the sport of our youths. We've discussed all other things, from motivation and faith to traveling and politics. He has a very matter-of-fact opinion on it all, making him something of a dream interview for a journalist. He doesn't do bullshit, and that's not uncommon among Norsemen, no matter how little time they spend in Norway.

The downpour turned to a drizzle, so I had a small window to walk briskly to the train station less than half a mile away. Kristian had graciously offered to walk me there, even though I assured him his legs are far too important to waste steps on me. He insisted, and it offered me the opportunity for one last question, so I asked him the one I try to ask every athlete and typically save for last: What would you be doing with your life if you weren't doing what you're doing?

This one requires a bit more thought, and I let myself feel lofty for a moment because I've asked the man I believe is the fittest human on earth a question he's never been confronted with. When he was ready, he was as certain as always: "If I weren't a professional athlete, I'd probably be working at a bike shop or for a bike company. But I think I'd always try to be a professional athlete, so I'd probably try to make it as a cyclist on the World Tour."

I was a bit taken aback because his background is swimming, and the run is where he wins his races, but a few moments later, it began to make sense. During our run up Mount Fløyen the day before, Kristian stopped to note the finish line of the 2017 UCI Road Cycling World Championship, which finished with the legendary Peter Sagan of Slovakia inches ahead of Alexander Kristoff, the greatest Norwegian cyclist in history.

From there the conversation shifted to Tadej Pogačar, who was about to win his second Tour de France. There's a reverence as he talked about Pogačar—who is nearly five years younger than Kristian. He's fascinated with how someone so young could be so much better than the rest—at a sport he thinks may be the toughest on earth. It's something Kristian is familiar with, but the awe with which he speaks of Pogačar is a deviation from his mean. I'd asked him about a lot of triathletes over the previous couple of days, and he'd mostly brushed them off. He believes—and knows—he's better than all of them. (Except, sometimes, his best friend.)

But the top tier of cycling is different, and perhaps that's why he flirted with the idea of switching sports. But, at least for now, he plans to compete in his fourth Olympic triathlon in Los Angeles and win a medal no one thinks he can win. When you get to the point that finishing an Ironman in under eight hours is relatively easy, it takes something like an impossible goal to turn you on. I'm assuming. Finishing an Ironman in any amount of time will never be relatively easy for me, but in researching and writing this book and adopting more Norwegian training principles, I'm finding that I can be a better runner in my 40s than I was in my 20s, and there's a lot more building up than breaking down ahead.

And that's all we're really trying to do. Get better and keep building. Unless you're as ambitious as the athletes featured in this book, there's no real rush. You may not be getting any younger, but pushing your body harder than it wants to be pushed—at any age—will eventually cause it to break down. The reason the athletes featured in this book are able to have such elevated aspirations is that they began building a massive base as early as possible. Now they have the extraordinary foundation needed to make extraordinary risks possible.

For Kristian, the next remarkable risk hangs in the balance. He's hinted at his ambition to race on cycling's biggest stages, but a fourth Olympics in triathlon and one more shot at a medal may be even more alluring.

Should Kristian make a departure from long-course triathlon, the Ironman realm will be Gustav's to rule. Meanwhile, Olav Bu, their shared coach, might eventually see his version of the method he's spent the last decade crafting come to conquer two sports at once.

For Arild Tveiten, the proof of his version of his method is in the extended boom of Triathlon Norway, which sent two women and a relay team to the Olympics for the first time in 2024. Only three Olympic cycles into its existence, "the little federation that could" continues to defy all odds.

For Ingrid Kristiansen and Marius Bakken—the de facto parents of the Norwegian method—they already have all the proof they'll ever need. They both know they got the absolute most out of themselves as athletes and their legacy lives on in the runners they coach and inspire today.

And for Jakob Ingebrigtsen, the runner carrying the torch of the Norwegian method—for now—the proof of his version of the method will be where it takes him next. A second consecutive gold medal in the 5000 in Paris proves that the method designed specifically for the 5000 works perfectly for exactly that. But becoming the greatest runner of all time will require much more than just being the greatest ever at one event. The Norwegian method is very much Jakob's to own and mold for the time being, and its value—at least to the outside world—will be judged largely on his results.

Within Norway, the merit of the method is evident as Norwegian athletes continue to succeed in the endurance world and beyond. When 22-year-old Markus Rooth won the decathlon at the 2024 Olympics, it was a surprise to many, but not the Norwegian Federation. The decathlon was just another box to tick on the quest to amass medals in the summer sports as they are known to do in winter. After all, no matter winter, summer, or something in between, getting outside and sustaining momentum is the most Norwegian thing one can do.

There's a Norwegian concept that has gained popularity in recent years known as *dørstokmilla*. It literally translates as "doorstep mile," and it refers to the idea that the hardest part of any endeavor is the step that gets you out the door. Basically, it's much harder to get moving than it is to sustain movement. That's true when it comes to starting big goals like qualifying for the Boston Marathon or Hawaii Ironman, or the shortest-term goal—starting your next session.

Perhaps the real secret of the Norwegian method is in getting out the door early, both in life and each day: Get up, get moving no matter the conditions, sustain momentum, and don't take too many days off. And, like a Viking, don't be afraid to shed a little blood.

APPENDIX

The Big Picture: How Endurance Terminology Stacks Up

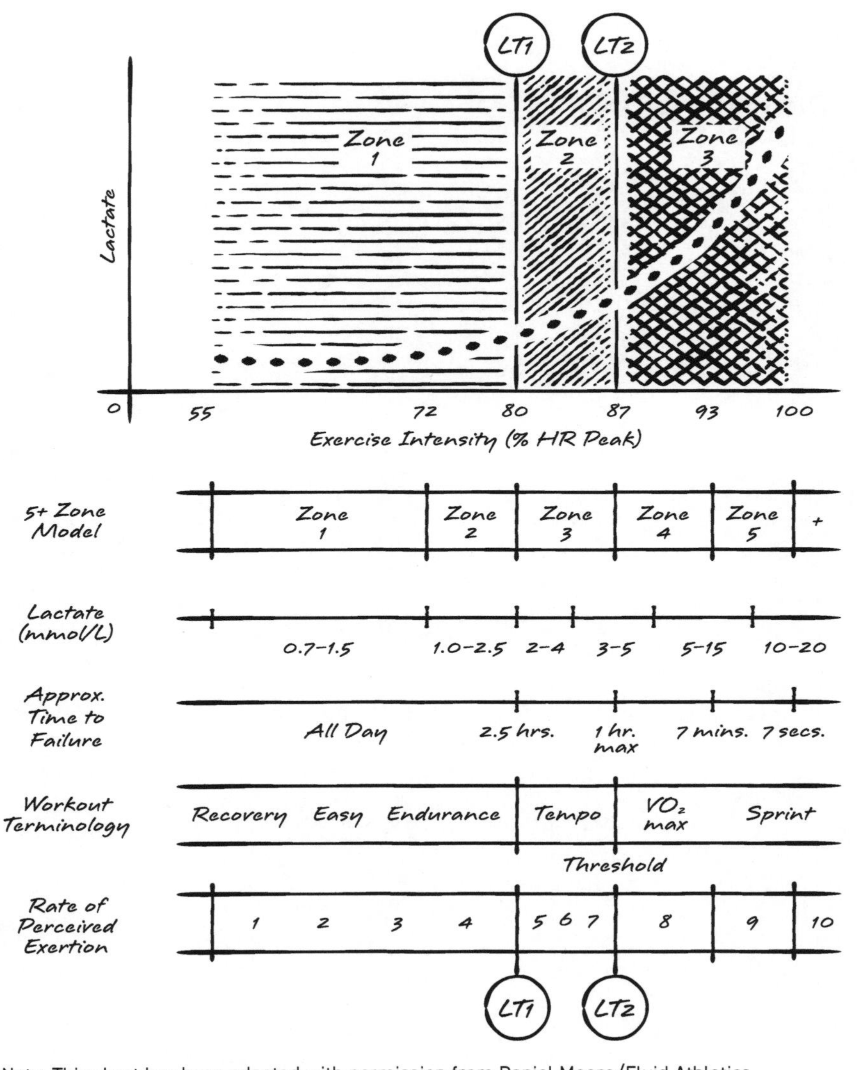

Note: This chart has been adapted with permission from Daniel Moore/Fluid Athletics.

ACKNOWLEDGMENTS

Thanks to everyone in Norway who made this book possible, in particular the four coaches who have been so crucial in the development and evolution of the Norwegian method. Thank you, Ingrid, Arild, Olav, and Marius, for your openness and for responding to dozens of emails and requests for calls from an American journalist who was very curious about your country and methods. I will miss our calls and the brief glimpses into the various corners of your beautiful country.

Thanks to Kristian for showing me around your hometown, even if I like it a bit more than you. I think you really should spend more time at home and hire a landscaper, but I get that the weather, riding, and food are just a bit better in France and Spain. Thanks for being a continued source of motivation for my own training. You will probably never get as many social media followers as Jakob, but your relentless pursuit of peak human performance will continue to inspire athletes from around the world.

Thanks to Siren Seiler, who was my mole into the world of toppidrett skoles, Olympiatoppen, and the Norwegian School of Sports Science. And thanks for getting responses from your dad much better and faster than I ever could. Thank you to Dr. Stephen Seiler, who, like me, was once an American very curious about Norwegian sports science. He is now an adopted and honorary Norwegian, and his research and wisdom are found throughout this book.

Thanks to my adopted hometown of Geneva, Illinois, for its strange fascination with Scandinavian culture, which led to my own. Thanks especially to the public library, where most of this book was researched and written. Find yourself a library with an entire Scandinavian wing and a piece of a Viking longship that will also entertain your kids for a bit. If not, just support your local library, anyway.

And thanks to my wife and kids for letting me work on this with morning and night double-threshold sessions so that most days can be reserved for long voyages to plunder carbohydrates.

Tusen takk! (Thousand thanks!)

RESOURCES

Chapter 2, Viking ship

Learn more about the Gokstad replica that sailed from Norway to New Orleans in 1893, including a short documentary, courtesy of Friends of the Viking Ship:

https://vikingship.us

Chapter 3, Ingrid Kristiansen

Visit the *Sports Illustrated* vault to read legendary American runner Kenny Moore's profile of Ingrid Kristiansen:

Kenny Moore. "The Best Norse in the Long Run." *Sports Illustrated*, Oct. 27, 1986.

Chapter 5, Arild Tveiten

Get Arild Tvieten's take on the science of triathlon:

Mikhail Eriksson. Interview with Arild Tveiten and Casper Stornes on triathlon training the Norwegian way. *That Triathlon Show*. Podcast audio. Episode 223: Mar. 2, 2020. https://scientifictriathlon.com/tts223/

Chapter 6, Norwegian Frequency Project for weightlifting

For a deeper dive into the Norwegian Frequency Project, check out this complete overview by Stronger by Science:

Greg Nuckols with Martijn Koevoets. "High Frequency Training for a Bigger Total: Research on Highly Trained Norwegian Powerlifters." *Stronger by Science.* https://www.strongerbyscience.com/high-frequency-training-for-a-bigger-total-research-on-highly-trained-norwegian-powerlifters/

Original study, delivered as a conference paper:

T. Raastad, A. Kirketeig, D. Wolf, G. Paulsen. "Powerlifters Improved Strength and Muscular Adaptations to a Greater Extent When Equal Total Training Volume Was Divided into 6 Compared to 3 Training Sessions per Week" (abstract). Book of abstracts, 17th annual conference of the ECSS, Brugge. Jul. 4–7, 2012.

Chapter 6, 8, 16, Dr. Stephen Seiler

Seiler followed sports science to Norway and now you can follow his research and insights into what the Norwegian method is (and isn't) on X:

https://x.com/StephenSeiler

Chapter 7, The blog that began a revolution

Marius Bakken synthesized everything he knew about the Norwegian method—which he called the Norwegian model—on his blog in the early 2000s. He has updated it over the years, including his thoughts on Jakob Ingebrigtsen, Kristian Blummenfelt, and Gustav Iden:

Marius Bakken. "The Norwegian Model: A Practical Guide." March 2000. http://www.mariusbakken.com/the-norwegian-model.html

Chapter 8, High-altitude rowing research

The UCLA Bionics Lab has published Dr Seiler's entire longitudinal altitude study on Norwegian rowers originally published in the *Scandinavian Journal of Medicine and Science in Sports* in 2004:

A. Fiskerstrand, K. S. Seiler. "Training and performance characteristics among Norwegian International Rowers 1970–2001." *Scandinavian Journal of Medicine and Science in Sports* 14(5): 303–310. http://bionics.seas.ucla.edu/education/Rowing/Training_2004_01.pdf

Chapter 8, 25 years of LHTL

Human Kinetics has the full text and PDF of the University of Western Australia's 25-year longitudinal study of the effects of "live high, train low":

Olivier Girard, Benjamin D. Levine, Robert F. Chapman, and Randall Wilber. "'Living High-Training Low' for Olympic Medal Performance: What Have We Learned 25 Years After Implementation?" *International Journal of Sports Physiology and Performance* 18(6): 563–572. https://journals.humankinetics.com/view/journals/ijspp/18/6/article-p563.xml

Chapter 9, Norwegian Method podcast

To keep up with Olav Aleksander Bu and learn more about the training methods he is currently exploring, listen to his podcast with Dr. David Lipman (and frequent guests Kristian and Gustav):

https://podcasts.apple.com/gb/podcast/the-norwegian-method-podcast/id1724540879

Chapter 10, Human Flourishing Index

The 2020 *Lancet* study that named Norway the best place in the world to raise children can be found at:

"A Future for the World's Children? A WHO–UNICEF–*Lancet* Commission." *Lancet* 395(10224): 537–658. Feb. 22, 2020. https://www.thelancet.com/journals/lancet/article/PIIS0140-6736(19)32540-1/fulltext

Chapter 13, More from Doha

Read Alex Hutchinson's in-depth review of the 2016 Doha heat study on cyclists:

Alex Hutchinson. "How Elite Athletes Respond to Extreme Heat." *Outside*. Dec. 8, 2018. https://www.outsideonline.com/health/training-performance/how-elite-athletes-respond-extreme-heat/

Original study:

S. Racinais et al. "Core Temperature Up to 41.5°C During the UCI Road Cycling World Championships in the Heat." *British Journal of Sports Medicine* 53(7): 426-429. https://pubmed.ncbi.nlm.nih.gov/30504486/

ABOUT THE AUTHOR

Brad Culp is a sports journalist, contributing to *Red Bulletin*, *Triathlete*, and just about every other triathlon publication on earth. He's worked as editor-in-chief of *Triathlete* and *LAVA* magazine, formerly the official publication of the Ironman triathlon series. He also served as media manager for World Triathlon for the 2010 season. This is his first book.

Growing up in suburban Chicago, his love of endurance began in the pool at age four. He earned All-American swimming honors at Fenwick High School in Oak Park, Illinois, where he also won three water polo state titles and had a hand in one. Culp finished his first Ironman triathlon during his junior year of college, where he was the youngest athlete competing at Ironman Wisconsin at age 19. He founded the Miami University (Ohio) Triathlon Club in addition to playing water polo.

Culp has finished more than 100 races and has DNF'd exactly one. He's done 7 Ironman triathlons, 15 marathons, and the Alpe d'Huez Triathlon in France, which is his favorite race on earth—and it's not particularly close. He hopes to change that by doing Norway's Norseman Xtreme Triathlon as soon as possible.

He lives with his wife and two children in Geneva, Illinois, hometown of two triathlon Olympians (so far) and one of the best places in the Midwest to be an endurance athlete. He looks forward to following the Norwegian method a bit more precisely when his kids are older and he gets over his fear of pricking his ear with a tiny needle.